MARKETING HIGH TECHNOLOGY SERVICES

MARKETING HIGH TECHNOLOGY SERVICES

Colin V. Sowter
MA (Oxon), D.Phil

Routledge
Taylor & Francis Group

LONDON AND NEW YORK

First edition published by McGraw-Hill Book Company Europe.

Published 2000 by Gower Publishing

Reissued 2018 by Routledge
2 Park Square, Milton Park, Abingdon, Oxon OX14 4RN
711 Third Avenue, New York, NY 10017, USA

Routledge is an imprint of the Taylor & Francis Group, an informa business

Typeset in 10 point Garamond Light by Acorn Bookwork, Salisbury, Wiltshire.

A Library of Congress record exists under LC control number: 00024870

ISBN 13: 978-1-138-72538-6 (hbk)
ISBN 13: 978-1-138-72536-2 (pbk)
ISBN 13: 978-1-315-19190-4 (ebk)

CONTENTS

List of figures   ix
Preface   xi
Acknowledgements   xv

PART I   THE ROLE OF MARKETING   1

1. What Is Marketing?
   A Way of Thinking about Business   3

Key business issues — 1.1 Marketing should be a means of
generating profit — 1.2 Marketing is a rational process that
should not compromise our integrity — 1.3 Marketing does not
always require large budgets or extra staff — 1.4 'Better
mousetraps' do not work; 'better doors' do! — 1.5 Becoming
market led involves a shift of focus — 1.6 Nobody actually
wants what we do – they want what it will do for them —
1.7 Marketing starts with identifying customer needs —
1.8 Marketing is too important to be left to the Marketing
Department — 1.9 Professionalism in marketing should be
recognized and sought — 1.10 Marketing addresses three
components of profit — 1.11 Marketing and selling have a
different focus — 1.12 What marketing is not! — 1.13 Being
market led does not mean satisfying every whim of every
customer — Exercises — Follow-up

2. Why Do People Buy?
Value as Perceived by the Customer   23
Key business issues — 2.1 The marketing mix — 2.2 The role
of price in the buying decision — 2.3 Customers buy a whole
bundle of attributes — 2.4 The features and benefits technique
and its limitations — 2.5 Different market segments want
different benefits — 2.6 Each member of a decision-making
group wants different benefits — 2.7 The unique selling
proposition — 2.8 The sharper cutting edge — 2.9 Branding
— 2.10 Some differences between intangible services and
tangible products — Exercises — Follow-up

3. What Is Involved in Becoming Market Led?
Setting our Sights High   47
Key business issues — 3.1 The development of the market-led
philosophy — 3.2 Culture changes involved in becoming
market led — 3.3 The benefits of a strong marketing function
— 3.4 Maximizing profit from existing resources — 3.5
Compatibility of targets and resources — Exercises — Follow-
up

4. To Whom Are We Selling?
Market Definition, Segmentation and Targeting   63
Key business issues — 4.1 What business are we in? —
4.2 Market segmentation — 4.3 Segmentation criteria —
4.4 Strategic decisions arising from market segmentation —
4.5 Marketing actions arising from segmentation — 4.6
Targeting — 4.7 The strategic significance of niche markets —
4.8 Credibility and track record — Exercises — Follow-up

5. Who Are our Competitors?
Competitive Analysis and Tactics   83
Key business issues — 5.1 The nature of competition —
5.2 Analysis of competition — 5.3 Sources of information about
competition — 5.4 Competitive tactics — Exercises — Follow-
up

6. How Do We Set Prices?
An Under-Exploited Management Opportunity   93
Key business issues — 6.1 Value pricing – price is what the
market will bear — 6.2 Customers do not know or care about
our cost — 6.3 Different parts of the business have different

profit potential — 6.4 There is no such thing as 'true cost' —
6.5 The danger of a commodity price orientation — 6.6 How
do we determine the market price? — 6.7 Pricing should be
used both strategically and tactically — Exercises — Follow-up

PART II   THE ROLE OF MARKETING COMMUNICATION   113

7. **How Do We Communicate?**
   Principles of Communication   115
   Key business issues — 7.1 The communications process —
   7.2 'Selling the next step' — 7.3 The role of different methods
   of communication — 7.4 The role of external agencies —
   Exercises — Follow-up

8. **How Do We Sell?**
   Personal Communication   127
   Key business issues — 8.1 Selling can be learned — 8.2 Selling
   starts with listening rather than speaking — 8.3 Presentations
   should be prepared and targeted — 8.4 Overcoming objections
   — 8.5 Closing the sale — 8.6 The selling role of non-sales staff
   — 8.7 Nurturing existing customers — 8.8 Partnerships and
   strategic alliances — Exercises — Follow-up

9. **How Do We Use Impersonal Communications?**
   Literature, Proposals, Web Sites, Letters, Mailshots, E-Mail,
   Telephone, Exhibitions, Advertising and Public Relations   143
   Key business issues — 9.1 Promotional literature —
   9.2 Proposals and invitations to tender — 9.3 Web sites —
   9.4 Sales letters, mailshots and e-mail — 9.5 The telephone —
   9.6 Exhibitions — 9.7 Advertising — 9.8 Public relations —
   Exercises — Follow-up

PART III   THE ROLE OF BUSINESS DEVELOPMENT   171

10. **What Does the Market Want?**
    Market Research, Marketing Research   173
    Key business issues — 10.1 Business decisions depend upon
    knowledge of the market — 10.2 Qualitative market research
    — 10.3 Quantitative market research — 10.4 Market research
    on a low budget — 10.5 Marketing research — Exercises —
    Follow-up

11. **How Do We Achieve Profitable Innovation?**
    Market-Led Innovation   189
    Key business issues — 11.1 Options for business development
    — 11.2 The role of marketing in innovation — 11.3 Make
    decisions before spending money — 11.4 Identify winners as
    early as possible — 11.5 Risk and sensitivity — 11.6 Dealing
    with impending failure — Exercises — Follow-up — Appendix:
    Discounted cash flow

12. **How Do We Manage the Future?**
    Market-Based Business Planning   205
    Key business issues — 12.1 The basis of business planning —
    12.2 Organization of the business development function —
    12.3 The planning process — 12.4 The business plan —
    12.5 The marketing plan — 12.6 The sales plan — 12.7
    Successful internal proposals — 12.8 Models and techniques —
    Exercises — Follow-up

    Templates for the Business Plan   231

    Index   243

LIST OF FIGURES

2.1	Creating value in the mind of the customer	28
2.2	The sharper cutting edge	39
3.1	The strong–weak interaction	56
3.2	Maximizing profit from existing resources	58
4.1	Industry segmentation	67
4.2	Sequential segmentation	69
4.3	Niche markets	75
6.1	The price–volume relationship	106
6.2	A saturated market	107
6.3	A segmented market	108
6.4	The effect of price on perceived value	108
7.1	The role of personal and impersonal communications	122
7.2	Size and diversity of target market	123
9.1	Reply form	156
10.1	Management style	175
12.1	Skill-set versus market segment organization	210
12.2	SWOT analysis	227
12.3	The Growth Share Matrix	228

PREFACE

HIGH TECHNOLOGY SERVICES DESERVE HIGH-QUALITY MARKETING

We live in a very competitive market economy. Whether we like it or not, commercial success often comes to those who are better at marketing rather than to those with better technology. Our technological skills need to be complemented by marketing and business skills.

Written by a scientist for scientists, the book is aimed partly at those working full time in marketing or business development departments in high technology service companies, but it is equally valuable for the increasing numbers of scientists, engineers, consultants and other professionals who have to market their own services in organizations with no specialist marketing or business development function. In my experience, even where such a function does exist, the full-time marketing staff are the first to say that they cannot do it all; they need the combined efforts of the whole team. Marketing is too important to be left to the Marketing Department – especially if there isn't one!

Readers may be working in scientific research, consultancy, development, project management, software, laboratory testing, engineering services, training, expert witness and countless other types of high technology services. They may be directors, managers or fee-earners. They may work in large companies, small companies or on their own. They may be in organizations which were not formerly regarded as businesses, such as government laboratories, research departments or other internal service providers; usually they are being told that the 'owners' are free to look to outside competitors, and the service providers are required to generate income from third parties.

An important theme developed in the book is that there is profitable business out there for those who have the skills and take the trouble to find it; we cannot wait for clients to come to us. If, as a result of reading the book, readers can win one more contract or justify a higher price in one case, the investment will have paid for itself many times over.

THE AUTHOR

I am not an academic, and this is not a theoretical study of marketing. It represents the fruit of 40 years' experience in a wide variety of companies. The lessons come from seeing marketing performed well and badly under UK, American and continental management styles. This has involved working in seven marketing positions, three of them at board level; wrestling with the role of marketing in the organization from a general management perspective; marketing the commercial services of the technological faculties of a university; running programmes for scientists and engineers in which they learn the basics of marketing; helping such people to prepare and present business plans to senior management in order to justify their very existence in the organization. For the last ten years I have had to market my own professional services in a highly competitive field. It is through these experiences that I have learned the lessons that I am seeking to share through the medium of *Marketing High Technology Services*.

This book is based on a previous edition, *Marketing for the Non-Marketing Manager*, originally published by McGraw-Hill but now out of print. The text has been specifically adapted to relate to the marketing of high technology services.

A MISUNDERSTANDING ABOUT MARKETING

To many people, marketing is little more than communication. They receive a wide variety of communications – by post, telephone, e-mail, fax, television, advertising and so on. This must be what marketing is about, and therefore they had better start doing some of these things!

The danger is that they confuse the *vehicle* with the *message* that the vehicle is intended to communicate. There is no point in having an expensive vehicle if it is carrying the wrong goods! People assume that if they pay enough to have an advertisement written or a brochure designed, all will be well. What they forget is that they are probably the only people who can create the right message by thoroughly understanding the needs of the client and the way those needs can be met. This is particularly true when what we are offering is an intangible service. There may well be a role for outside experts in communication, but we dare not delegate the creation of the message to people who are not able to understand it.

Creating the right message is not something that necessarily comes instinctively to high technologists. We are trained to understand our sophisticated technology in great detail. Faced with the need to communicate, we tend to tell people *who we are and what we do*. Although this is very common, it is not good marketing; it is an 'in-to-out' process, which starts with us and what we have to offer. Marketing thinking should be 'out-to-in', starting with the needs of the customer and working back to how we can meet that need.

THE STRUCTURE OF THE BOOK

We cannot start thinking about communicating until we have answered some key questions. What audience do we want to address? Do they need our services, and do they know that they need them? What are the main factors which will influence their buying decisions? Who is involved in making the buying decision; if it is a multi-level multi-discipline decision, are we in contact with the right people? What is the competitive environment within which we are working? What price are they prepared to pay, and are they prepared to pay a higher price for a more valuable service?

Once we have considered these fundamental business issues in Part I of this book, we move on to think about the various forms of communication in Part II. The processes, which culminate in a number of checklists, are virtually guaranteed to make our communications more effective.

Part III looks at market research leading on to the crucial subject of innovation. Why is it that high technology innovation is so ineffective in many organizations? Having been part of the team that moved into newly acquired companies on several occasions, I have probably closed down more development projects than I have opened. This book argues that innovation does not take place solely or even mainly in the Research and Development (R&D) Department; the marketing component of innovation is at least as important.

Finally, we consider some aspects of marketing management which are often badly handled, focusing particularly on the preparation of business plans. These are not theoretical documents; they are designed to show whether or not we have a viable business, and are a means of convincing us, our senior managers and sometimes external providers of funds that what we propose to do offers a sound basis for investment. They are founded on external marketplace realities, not on internal aspirations (another example of 'out-to-in' thinking). This new section is based on a series of templates which can be used as the basis of a business plan.

As with the seminars on which it is based, this book is intended to be strongly action-oriented. The issue throughout is 'if we agree with it, what are we going to do about it?' For this reason, each chapter ends with some very practical 'Exercises' and a 'Follow-up' section. These are noted for reference at the appropriate points in the text. Most of them will be even more useful if they are worked on with a colleague or a small group.

AUTHOR'S CONTACT ADDRESS

I would be happy to receive comments on any aspects of the material in this book, or to discuss its application. Please contact me in the first instance by fax on

01483 892894 from the UK,
+ 44 1483 892894 from outside the UK

or by e-mail on sowter@lineone.net

Colin Sowter

ACKNOWLEDGEMENTS

While no material has knowingly been copied from other sources without acknowledgement, I am indebted to the large number of authors whose books I have read over the years, the consultants whose seminars I have attended, the directors and managers under whom and with whom I have worked in various appointments in a variety of industries, and the delegates who have participated in my own seminars.

In particular, I would like to thank Frans Waals for his wise and sensitive advice during the course of a major business planning operation in which we were both involved.

The combined wisdom of all these people has led to the experience that is the subject of this book, and I would like to acknowledge my gratitude to each one of them.

CVS

PART I
THE ROLE OF
MARKETING

❖

1

WHAT IS MARKETING?
A WAY OF THINKING ABOUT BUSINESS

Key business issues   *Section*

O Marketing should be seen as a means of generating
   profit. If it is not, there is something wrong, and the
   problem needs to be resolved   1.1

O Marketing is a rational process which should not
   compromise our intellectual, professional, technical or
   ethical integrity   1.2

O Marketing does not always require large budgets or extra
   staff – it is a way of thinking about business   1.3

O We cannot assume that people know or want what we
   sell, or that they are prepared to find out   1.4

O Marketing should be proactive not reactive

O The 'best' service is not always the most commercially
   successful

O Becoming market-led involves a shift of focus away from
   'what we do' to 'what they need'   1.5

○ Nobody, not even our best customer, actually wants what
  we do; they want what it will do for them   1.6

○ Marketing starts with identifying customer needs, and then
  organizes the whole company operation to satisfy those
  needs in a profitable way   1.7

○ Marketing is too important to be left to the Marketing
  Department. Key staff in all functions must be market led   1.8

○ Professionalism in marketing should be recognized and
  sought. Many senior managers set their sights too low in
  recruitment and appointment   1.9

○ Marketing influences three components of profit – market
  share, market size and margin. Two of these are often
  ignored   1.10

○ Marketing and selling have an entirely different focus and
  should not be confused   1.11

○ Marketing is *not* telling people what we do   1.12

○ Marketing is *not* persuading people to buy what they do
  not need

○ Marketing is *not* only about brochures, advertising,
  customer support or customer service

○ Being market led does not mean satisfying every whim of
  every customer   1.13

1.1   MARKETING SHOULD BE A MEANS OF GENERATING PROFIT

'The trouble with marketing is that it is a debit against profit.' This remark, made to me at one of my seminars, reveals volumes about some people's attitudes to marketing. It suggests that it is some sort of disease that needs to be eradicated! Unfortunately this is a commonly held view, and it is based on a complete misunderstanding of what marketing is there to achieve.

Most of what we do in business is a debit against profit, and marketing is no exception. However, if it is not a means of *generating* profit there is something wrong, either with what we are selling, the market demand or the way in which we are setting about marketing. The problem needs to be identified and dealt with, and this may involve a change in the whole operating philosophy of the company. The fact is that, in a market-led operation, the money spent on marketing is seen as an investment from which we expect a positive return. The short-term negative profit and cash flow is the price we have to pay for long-term success.

Inept and inappropriate marketing is certainly a debit against profit!

Sadly, people venture tentatively into some form of marketing that they have seen elsewhere, are disappointed with the results and conclude that marketing does not work in their particular business. It is not the fault of marketing but of the way in which they are trying to apply it. Wherever there is competition and customers do not spontaneously demand our service, marketing is necessary. It is not enough to offer a good service – excellence is expected. Whether we like it or not, the actual service we are selling may not be the main factor in the buying decision, and success goes to those who are best at marketing.

This is equally true whether the 'product' is a tangible item such as a computer, or something intangible such as a professional service.

We are not suggesting that money should be poured aimlessly into a bottomless marketing pit. But, intelligently applied, time and money spent on marketing can be the biggest single factor contributing to profitable growth (Exercise 1.1).

1.2 MARKETING IS A RATIONAL PROCESS THAT SHOULD NOT COMPROMISE OUR INTEGRITY

Until relatively recently, one could obtain qualifications, degrees, further degrees and so on in almost every subject except marketing. A consequence of this is the widespread feeling that marketing is not quite as 'professional' as other disciplines. Doubts about some of the highly emotive activities such as advertising and promotion even lead some people to regard marketing as a rather inferior and irrational process.

The aim of this book is to demonstrate that, properly applied to each particular set of circumstances, marketing should not compromise our intellectual, professional, technical or ethical integrity in any way. The attempt to persuade people to buy products or services that they do not really need might be described as marketing, but it is not the subject of this book. It is wrong to indulge in irrational or dubious marketing, and no reputable business would do so.

1.3   MARKETING DOES NOT ALWAYS REQUIRE LARGE BUDGETS OR EXTRA STAFF

Marketing is an attitude of mind that can transform the way existing staff carry out their normal tasks. It does not always require additional staff or large budgets. The people who have to deliver the service are often the best people to market it. They have an unrivalled knowledge of the needs of the customer and the way in which they can contribute to meeting those needs.

Where the rewards for success are very high, leading companies spend a large percentage of turnover on marketing in its various forms – market research, marketing strategy, market-led innovation, selling, sales support, customer service and so on. This makes good business sense. If millions of pounds are to be spent on the development of a new skill or service, it is essential to establish potential customer demand before committing to the development (but how many companies do this?). If a large sales force is to be employed, their activities must be prioritized on the basis of marketplace needs. If large budgets are to be spent on television advertising, it is worth commissioning substantial market research to test the advertisements before committing to the campaign.

While this book is relevant to these high-spending businesses, we particularly deal with the situation where a large budget is not available and would not be appropriate. Many of the delegates to my seminars come from organizations that do not even have a marketing department. For many of us, testing everything by formal market research is simply not realistic, and we have to make judgements in the absence of information that we would like to have. This is one of the reasons why a marketing management post demands abilities of the highest calibre.

Marketing is not primarily a series of separate, specialist and costly activities; it is a way of running a business, a management style, an attitude of mind. As such it need not, in the first instance, cost anything. The salaries and expenses of the staff concerned are already being paid. A market-oriented management style should saturate every part of our daily decision-making process, so that the main factors influencing strategic and tactical decisions are the needs of the customer rather than the needs of the company.

Many companies which have survived with very low levels of marketing expenditure would achieve greater profitability and growth if they were prepared to take marketing more seriously. In many cases, the operation has been the internal reactive service provider of a larger business, accounted for as a cost centre; now it is seen as a profit centre, required to justify its existence and to generate its own funding in a competitive free market. This will undoubtedly involve spending more money on appropriate aspects of marketing, and of course there is a time lag between investment and return, but this is true of many aspects of business and is not a valid reason for not engaging in marketing.

1.4 'BETTER MOUSETRAPS' DO NOT WORK; 'BETTER DOORS' DO!

> If a man ... make a *better mouse-trap* than his neighbour, tho' he build his house in the woods, the world will make a beaten path to his door.
>
> (Attributed to R. W. Emerson by S. S. B. Yule in her *Borrowings*, 1889, emphasis added)

This statement is often seen on the walls of company entrance halls or development departments. It contains a number of dangerous fallacies.

First, we cannot assume that people know what services we provide. Many companies base their operating procedures on a philosophy of 'enquiries received'. It is much healthier to assume we shall not receive a single enquiry that we ourselves have not generated. Many companies are diversifying out of their historic core business into new fields where they are unknown. Their name would not be on even a 'long list' of potential suppliers to this marketplace, let alone a 'short list'; it is the task of marketing to overcome this.

Second, we cannot assume that people want what we are selling, in the sense that they already have it on their 'shopping list'. A prime task of Marketing is to stimulate or generate a demand that may be only latent or potential.

Third, we cannot assume that people will 'beat a path to our door'. We need to be proactive, not reactive. 'Order-taking' has little place in a positive marketing strategy. Of course we will take orders when they are presented to us, but the emphasis should be to go out, find potential customers, identify their needs, demonstrate that we have what they need and generate the contracts by our own initiatives.

Fourth, there is the question of whether people want 'better mousetraps'. Superficial thought suggests that the more sophisticated our service, the more likely we are to sell it. A little more thought suggests that people are not prepared to pay for 'gold plating', over-engineering or superfluous features. There are some thought-provoking acronyms to describe this quandary: JEE – 'just enough engineering', and AIP – 'adequate is perfect'. Different segments of the market (see Chapter 4) will want different degrees of sophistication. It is by no means obvious that the maximum profit will be achieved by addressing only the top tranche of a particular market. Culture problems can arise for people who have been trained in technical or professional excellence and feel that there is something inferior about the suggestion of 'compromise', but every sale is a balance between price and perceived value. The art is to pitch the level appropriately for any particular buying situation, giving the customer a product or service that is 'fit for purpose'.

In contrast to the quotation from Emerson, it has been said that 'If we make a better door, people will beat a path to our mousetrap'. This is a much better expression of marketing philosophy. In many companies, the need is not so much to improve the service as to improve the contact with the marketplace. Curiously, some managers seem to regard money spent on technical development as an investment but money spent on marketing as a waste (Exercise 1.2).

1.5 BECOMING MARKET LED INVOLVES A SHIFT OF FOCUS

Becoming a market-led operation requires a deliberate shift of focus, away from *what we do* to *what they need*, or want, or might want or might reasonably be persuaded to want, which is a problem for those who live in a culture of specifications, definitions, standards or professional methodologies. Scientific and professional education encourages us to seek perfection, to search for excellence and to devote all our attention to what we do. We must recognize that people's real needs do not necessarily coincide with our view of what we would like them to have. What the customer actually needs is a *business benefit*, not a technical benefit. They want increased profit, reduced cost, greater efficiency, the ability to serve their customers better, the opportunity to charge higher prices and so on. These business benefits may or may not be gained from a technical advance, but the technical benefit is never an end in itself; it is a means of obtaining the business benefit. We can understand the business benefits only if we understand the business.

Part of this shift of focus therefore involves learning the customers' language and jargon, identifying with the issues which concern them. It also requires a deliberate effort to suppress those parts of our own language and jargon which might be a 'turn-off'. This is particularly necessary when we are trying to diversify into a new area; we have to leave behind the old culture and learn to understand a completely new one.

Many companies pay excessive attention to their past achievements, extolling the length of time they have been in business, the fact that they were first in the field with a particular service and so on. The requirement from the customers' point of view is what the supplier is able to do for them in the future. Again, this may require a shift of focus.

As has already been said, and will be repeated many times in this book, we need to think 'out-to-in' rather than 'in-to-out'. This may have an impact, for example, on how we spend our time. A group of delegates told me that they spend over two hours a day dealing with their e-mails. If the majority of these are dealing with matters that really relate to the profitable development of their business, then fine! But if most of them are concerned with internal company issues with no relevance to the customer, we must question whether our priorities are correct.

1.6 NOBODY ACTUALLY WANTS WHAT WE DO – THEY WANT WHAT IT WILL DO FOR THEM

Nobody, not even our best customer, actually wants our skills or services. This sounds a drastic statement to make in a book about marketing, but it is worth thinking through its implications. What the buyers want is not the service itself but what the service will *do for them*. Developing new capabilities does not of itself achieve anything. These activities do not win the race; all they do is to get us to the starting point.

People do not want drills – they want holes. People do not want complex methodologies – they want a more efficient operation, less waste, less risk, better plant utilization, a better product for their customers, or whatever else they are seeking. A service is a means to an end, not an end in itself.

Failure to realize this can be seen in many of the brochures, advertisements, web sites, e-mails, sales claims and so on with which people are daily bombarded. The emphasis is far too much on telling them what the service does, and far too little on what benefits they can gain from it. This is why we have to start with our customers and their needs, not with ourselves and what we sell. As always, we must think 'out-to-in', not 'in-to-out'.

This means that we must be prepared to work at our marketing. It is easier and much less disrupting to live in our own world – we understand it and we are comfortable with its culture. We can comprehend and measure what we are doing, whereas the outside world is made up of unquantifiable mysteries. The fact is, however, that the prizes go to those who are prepared to launch out into the unknown world of the customer, with all the risks, frustrations and effort that this involves (Exercise 1.3).

1.7 MARKETING STARTS WITH IDENTIFYING CUSTOMER NEEDS

A 'definition' of marketing is that it starts with identifying customer needs and then goes on to organize the whole company operation to satisfy those needs in a profitable way. However, this is too simplistic in many cases.

First, the need may not be known to the customers themselves. There may be no spontaneous demand. We therefore have to start by creating the need. This does not mean that we are trying to deceive our customers into buying a service they do not require, but it does mean that we have to take the initiative in helping the customers to realize why they need us and what we have to offer.

Second, the need may be potential rather than existing. If we devote all our efforts to meeting present needs (most of which derive from past needs), we shall be followers rather than leaders and we shall miss many profitable opportunities which could be ours. For this reason, marketing must be concerned with our business in years to come before it can influence today's activities.

Third, the need may not be expressed; it may be unexpressed or latent, and it has to be developed in the mind of the customer. It may not have occurred to the customers to buy what we are selling – they may not even have heard of it. One of the prime tasks of the consultant is to help the customers to realize and then to articulate their own needs. We may have to sell a concept before we can begin to sell an actual service. In order to do this, we have to diagnose what the problems or opportunities actually are, and then develop the service packages that will solve those problems or take advantage of those opportunities. If we sit back and wait for people to send in enquiries, which is what many companies do, we will still be waiting.

The above comments have applied to most of the exciting growth areas of today. Customers did not take the initiative to demand the sophisticated applications of computers, information technology (IT), telecommunications, software, virtual reality, specialist areas of science and engineering and so on. People with vision and an understanding of what would benefit their customers took the risk, invested in the services, and then marketed them proactively and professionally.

The next part of the definition – 'organize the whole company operation to satisfy those needs in a profitable way' – often causes problems. It seems to imply that the Marketing Department is more important than other departments, and can tell the rest of the company what to do. This is a misunderstanding. All main departments in a company are necessary, just as a table needs all its legs.

The idea we are beginning to develop is that marketing does not take place solely or even primarily in the marketing department. It is a way of thinking about business, which must saturate management thinking in all departments and at all levels. It is this *marketing philosophy* which must be the prime input to company strategy, and this philosophy is too far-reaching to be confined to one department or one level in the management structure.

If the objective of marketing is to generate profit, does this refer to short-term profit or long-term profit, and do these two not conflict? The answer to these three questions is 'yes, yes and yes'. A company with a very short-term orientation is often very bad at marketing; managers are obsessed with making profit this year, this quarter or even this month, to the detriment of all else. Obviously there are times when we have to 'batten down the hatches and survive the storm', but short-term cost savings are not a formula for achieving long-term strategic growth.

In the start-up phase, successful businesses often invest a large proportion of their initial profit in marketing, in order to build a market position, create an image and buy a market share. Having done this, they are able to reduce the percentage expenditure on marketing and enjoy the sustained growth in profit that they deserve.

We are not suggesting that money should be poured unthinkingly into marketing activities. What we are advocating is a careful evaluation of the market opportunities, an informed estimate of the amount of business which can be obtained, a sensible assessment of the amount of marketing money required to win that business, a detailed comparison of the various ways in which the money might be spent (selling, advertising, literature, exhibitions etc.) and firm commitment to a positive marketing programme.

1.8 MARKETING IS TOO IMPORTANT TO BE LEFT TO THE MARKETING DEPARTMENT

A myth about marketing is that it is something that goes on in the marketing department. Well-dressed and highly paid individuals seem to carry out mysterious functions, use jargon understood only by themselves and take little interest in what goes on in the rest of the company. They regard themselves as a race apart from, and distinctly above, all the other functions. This leads some people to conclude that 'marketing does not work in our business' or 'we can't afford to employ people like that'.

Marketing should start with the chief executive, be a priority of the whole senior management team and percolate down to virtually everyone at senior and middle management level in the organization.

Apart from those who are required to be proactive in negotiating or helping to win contracts, there are many others who are involved in reactive marketing. They include a switchboard operator, a credit control clerk, a quality manager, a member of the legal department – in fact, anyone who has any contact with customers. These people may not be able to win a contract but they can certainly *lose* one!

In some organizations there is a designated function called 'Marketing', which is allocated certain specific tasks such as researching the market, developing strategy, preparing marketing programmes, selling to the marketplace and supporting customers. In others there is no separate function. In every case, the essential marketing tasks still have to be performed.

Marketing could be seen as:

O   customer oriented
O   financially sound
O   international
O   business management.

Marketing is *customer oriented* because the customer pays our salary. It is *financially sound* because what we are selling usually represents a financial benefit, either to our immediate customers or to their customers or both. Training in relevant financial concepts may need to be a part of training in marketing so that soundly based financial arguments can be presented.

Marketing is also *international* because, for most businesses, the potential from the home market is insufficient to meet corporate growth aspirations. With regard to *business management*, I was once asked to give a talk to a group of Marketing Managers in a 'blue-chip', fast-moving consumer goods company. My session was after lunch on the fourth day of a one-week course; the delegates were beginning to tire, and I feared that the unglamorous subject of 'Industrial Marketing' would be a complete turn-off. After the session, one of the delegates said: 'It's a good job *you* came to speak to us; we've been here for four days learning about gimmicks and techniques and jargon, but you're the first person to use the word "management" in connection with marketing.' Marketing is nothing less than a sophisticated form of business management.

Marketing is much too important to be left to the Marketing Department – especially if there isn't one (Exercise 1.4).

1.9 PROFESSIONALISM IN MARKETING SHOULD BE RECOGNIZED AND SOUGHT

Companies often correctly diagnose the need for a marketing function but fail to obtain the benefits because they set their sights far too low in terms of marketing professionalism. This is usually because senior managers have never seen marketing work the way it should. They often regard marketing as part of selling, and staff the marketing department with former sales people. In high technology companies, the most common mistake is to transfer a technical expert into the Marketing Department without any training, on the grounds that 'marketing is common sense'. The same company would not put a taxation specialist in charge of chemical engineering plant development; why should they think a comparable transfer will work in the case of marketing?

Marketing is now a profession that can be learned, studied and experienced. Where a company or an industry is not renowned for training heavy-weight marketing staff, it may well be worth bringing in at least a proportion of them from a company or industry that is. The immediate reaction to such a suggestion is often 'you can't understand our business until you have been in it for twenty years'. The reverse – bringing in a sense of objectivity which is not hampered by past practices and attitudes – might actually be just what is needed.

The flaw in making do with unprofessional and inexperienced marketing staff is that very little cost is saved. The salary may be less than one would have to pay for a professional, but all the other expenses – car, office, secre-tary, administration etc. – are exactly the same. One experienced professional can be worth several unqualified newcomers in terms of influence on the business.

In theory, the ideal recruit would have a knowledge of the marketplace, would be expert in the technological skills being marketed and would have had a good marketing track record. If such a person can be found, that is fine. However, candidates usually possess only one or two of these three qualifications. In these circumstances, which should be regarded as the most necessary? Although this is not the natural instinct of many managers involved in recruitment, it is suggested that the three factors should be ranked in the following order:

1.   Knowledge and experience of marketing in a sophisticated marketing environment.
2.   Knowledge of the marketplace.
3.   Expertise in the services being marketed.

Knowledge of the marketplace and the fundamental aspects of the services can be learned relatively quickly. Marketing people recruited from a totally different industry can adapt very quickly to the new situation. They have learned to ask the right questions and will largely train themselves. They will instinctively set about finding out the size of the market, the trends taking place, the activities of the competition, the reasons why people do and do not use the company's services and so on. I have seen such people chairing a meeting with great authority within two months of joining. Of course they are better after two years, but the alternative – appointing people who have never seen marketing properly used – is incomparably worse (Follow-up 1.1).

If this is so, does not this demolish the whole thesis of this book, which is that high technologists can play an essential role in marketing their services? Emphatically not! In some smaller organizations, it might not be appropriate to have qualified, full-time marketeers; experience has shown that open-minded high technologists can gain enough understanding of marketing from reading a book or attending a seminar to transform the way they approach their work, often with dramatic results and a greatly increased degree of job satisfaction. In larger organizations that do have a separate marketing or business development function, the full-timers invariably welcome the complementary marketing efforts of the fee-earners.

1.10 MARKETING ADDRESSES THREE COMPONENTS OF PROFIT

Marketing addresses three components which combine to make up the profitability of an enterprise. These are:

O   market share
O   market size
O   percentage margin.

The instinct of many people is to concentrate on the first component – market share. For them, the marketing battle is to compete for a limited amount of business and try to capture sales from the competition. They respond in an increasingly desperate spiral to invitations to tender, few of which they succeed in winning. While this is clearly an important part of marketing, it is only one part and greater rewards may actually come from the other two components.

MARKET SHARE

The starting point for any business plan or marketing plan is the sales forecast. The problem is that this forecast is often not expressed in terms of market share and it therefore lacks external validation. Any manager who is asked to approve a set of sales figures should insist on knowing at least approximately the market share which it represents (see Section 12.1).

Figures 'plucked out of the air' are a dangerous self-delusion and may be wildly optimistic. They are often based on the value of contracts that the fee-earning consultants need in order to keep them occupied. This is a sign of marketing at its most amateur. An annual growth rate of 10 per cent may be pessimistic if the market is growing by 15 per cent, but impossibly optimistic if the market is static or declining. In spite of this, immature managers feel that they must build growth into their forecasts irrespective of what is happening in the marketplace. The truth is often that they need this growth in order to justify the increases they are planning in their own empire. Similarly, senior managers often demand growth figures because they think these will create an incentive. Realistically stretching targets can be motivating, but targets that are not achievable have the reverse effect and simply defer the time when difficult management decisions have to be faced.

We must consider whether a market share figure is really achievable. If we are selling a new concept, how do we know that sales will 'take off' so quickly? If we are entering an established market dominated by other suppliers, why should the customers use this hitherto unknown service provider? Why should the competitors allow a new entrant to make these gains at their expense? Unless we can answer these questions, we should reconsider the sales forecast.

MARKET SIZE

The second component of profit – increasing the size of a market – may initially involve selling a concept rather than a service. For example, persuading people to call in an outside management consultant does not primarily involve proving that one firm of consultants is better than another; the task is to persuade people that they need outside help at all, and that this is a better option than struggling with under-equipped internal resources. A patent agent might carry out an audit on the way in which a client protects intellectual property; the result might be to create an increased demand for patenting. This is quite different from sitting back and responding reactively to enquiries received.

There are several ways of expanding the market for an existing service, such as broader application, new uses and under-exploited potential.

Some of the most dramatic case histories are those in which a whole new market has been created, which is larger than the original market that spawned it. Customers did not spontaneously demand environmental assessments, benchmarking programmes or safety consultancy, but entrepreneurs recognized and exploited the opportunities. This is sheer marketing creativity; the market is what we make of it!

PERCENTAGE MARGIN

The third component of profit – the percentage margin – should be a major concern of marketing. It is based on two elements – price and cost. In practice, pricing is often carried out by the Finance Department, and cost is regarded as something which arises mainly from salary scales, historic staffing levels and percentage utilization of fee-earners. The fact is that both price and cost are strongly related to the situation in the marketplace, and cannot be handled by people living in an 'ivory castle' which insulates them from the realities of customers and competitors.

Price is what the customer is prepared to pay for the value they perceive in what they receive; it has much less to do with cost than most people realize. This concept is so important that the whole of Chapter 6 is devoted to the subject. The impact of price on profitability is enormous. If our net profit is 10 per cent of turnover, an average price increase of 5 per cent would increase profit and the return on capital employed by 50 per cent!

Professional marketeers will try to justify a premium price by promoting the value to the customer of the service that they provide. They will be aware of the prices charged by the competition, which may have a significant influence on the price that people are prepared to pay for their service (but they still have to decide where to pitch their prices in relation to the wide band of prices charged by competitors). If this is true, prices cannot be set by accountants alone (unless they spend about three days a week in the marketplace!). It is, of course, assumed that the marketing staff are mature managers who understand the financial realities of business and who see their task as generating profit rather than turnover.

Cost is also strongly influenced by the marketing function. Of course, there are some aspects which are outside the control of marketing – company and corporate overheads, salary scales and so on – but marketing decisions can often mean the difference between profit or a loss. Marketing is responsible for defining the specification of the service package being offered, which has a direct impact on cost. On the negative side, incremental activities that the customer does not value will reduce the profit because the cost cannot be recovered in the price. On the positive side, a service with a higher specification may be a vehicle for increased market share or a higher price. A marketing specification is not a 'wish list'; it calls for a responsible appraisal of an infinite number of options, and a willingness to say 'no' to some interesting features.

These are the three key components of profit that are influenced by marketing, and we neglect any of them at our peril (Exercise 1.5).

1.11 MARKETING AND SELLING HAVE A DIFFERENT FOCUS

Words are used differently in different organizations, but the most common approach is to describe the whole discipline as marketing and then to subdivide aspects of sales and marketing within this discipline. The organizational implications of this are discussed in Section 12.2.

A sales emphasis is that we have a service that we need to sell. A marketing emphasis is that the market has a need that we can arrange to meet. The service is the result of the marketing effort rather than the reason for making the marketing effort. Excellence in selling is not sufficient if it is not accompanied by excellence in marketing.

SELLING

Selling should have the highly focused short-term objectives of winning today's contracts, achieving this month's sales forecasts and generating the income required to meet the annual plan. This activity is totally necessary but should not be seen as representing the whole of the marketing operation, as it appears to do in some companies.

STRATEGIC MARKETING AND MARKETING MANAGEMENT

Strategic marketing and marketing management are highly demanding activities which should be undertaken by staff of the necessary calibre and with the necessary experience. The task of the marketing strategist is to identify the long-term needs of the marketplace, to gain an informed view of the probable long-term actions of the competition and then to propose a strategy that will enable the company to exploit these opportunities profitably.

Arising from the overall marketing plans will be detailed plans for specific marketing programmes such as the development and launch of new services, updating and relaunching existing services, entry into new market segments, entry into new geographical areas, changes in pricing strategy, and all other activities that will enable the company to progress in a market-led manner.

Marketing is making the future happen!

MARKETING SERVICES

An ancillary aspect of the marketing function is to provide tools that will enable the sales force to achieve their objectives. These tools include literature, advertising, exhibitions, sales aids, sales manuals, training packages, service support packages and so on. This is the one area where marketing might legitimately be regarded as a support function to selling.

CALIBRE AND EXPERIENCE OF MARKETING STAFF

The various areas within the sales and marketing function clearly require very different types of experience, qualifications and motivation. Good marketing managers should be able to sell, and indeed should have had some experience 'on the road' to ensure that they are not abstract theoreticians, but selling would probably not provide sufficient job satisfaction in the long term.

Good sales staff do not necessarily transfer successfully to marketing positions. Their whole culture is based on the achievement of short-term results, and it is this that makes them successful. They may not be very good at longer-term or conceptual thinking, and they should not be expected to do it. They may have valuable insights into the marketplace and the competition, but it is probably better if the strategic thinking resulting from their input is carried out by staff with longer-term objectives. This is not to suggest that sales people are in any way inferior to marketing strategists; they are both necessary, but they have different strengths.

The advantages and disadvantages of having full-time sales staff, as distinct from relying on the fee-earners to sell their own services, is debated in Section 8.6. In all cases, the aim should be to recruit, train or appoint the right people for each task (Exercise 1.6).

In many service companies, particularly small ones, these functions are combined into one department or one person, or they may be part of one person's duties, which also include fee-earning. Such a function is often called 'business development', which sounds more respectable than 'marketing' and 'selling'. This is no problem, as long as the marketing and selling is achieved and everyone understands how responsibility is allocated.

1.12   WHAT MARKETING IS NOT!

While we are exploring the question of what marketing is, we should also understand what it is not. Many companies set their sights far too low in terms of what marketing is there to do. They seem to believe certain myths about marketing and thereby bring the whole concept into disrepute.

MYTH 1: MARKETING IS TELLING PEOPLE WHAT WE DO

This point has already been discussed, but it is included here as possibly the most widespread myth of all.

MYTH 2: MARKETING IS PERSUADING PEOPLE TO BUY THINGS THEY DO NOT NEED

People come to our companies and homes, uninvited and unwelcome, and try to persuade us to buy products or services against our will. Their only interest is in achieving a sale and gaining their commission. They probably only sell once in a lifetime to a particular customer, and they have nothing to lose by taking a very aggressive sales approach.

This type of so-called marketing is inappropriate to the high technology service business, where the aim is normally to build continuing customer relationships, based on trust and mutual benefit. Good marketing is a 'win-win' outcome.

MYTH 3: MARKETING IS ABOUT BROCHURES, ADVERTISING AND PUBLIC RELATIONS

These activities, although necessary, represent only the tip of the iceberg of marketing. The most important point is the message to be communicated. Without proper attention to the message, the money spent on communication is largely wasted. The correct message requires a great deal of time and thought and can only be achieved by staff with a clear understanding of the business needs of the customer and the way in which those needs can be met.

MYTH 4: MARKETING IS CUSTOMER SUPPORT AND CUSTOMER SERVICE

These activities are fundamental, but they are not a substitute for strategic marketing. In one company where this attitude had prevailed, I set up an entirely separate customer service function so that customers could be properly supported, but also so that marketing managers could get on with what they were there to do without constantly having to respond reactively to unplanned demands on their time.

MYTH 5: MARKETING IS ABOUT FASHIONABLE CLOTHES AND EXPENSIVE LUNCHES

Those who give this impression have only themselves to blame if their form of marketing is not taken seriously by hard-working managers in other functions whose only perk is an occasional company sandwich. Marketing is one of the most crucial disciplines in the company, and those who practise it should be ambassadors for the seriousness of their profession.

To summarize, senior marketing staff have the great privilege of pointing the company in the direction in which it should go. This privilege carries with it an enormous degree of responsibility. If they cannot meet the high demands placed upon them, they should be replaced by people who can (Exercise 1.7).

1.13   BEING MARKET LED DOES NOT MEAN SATISFYING EVERY WHIM OF EVERY CUSTOMER

So far we have urged that the needs of the marketplace should be the dominant factor influencing strategic and tactical business decision-making. Internal needs, such as wanting to keep our fee-earning consultants gainfully employed, must ultimately be subservient to the needs of customers. Our own desire to run a prosperous business depends critically upon having satisfied customers, at least in any situation where repeat purchases and customer loyalty are important.

These arguments seem to suggest that we should go to great lengths to do anything a customer asks us to. This is not the case, and an attempt to do so will lead to disaster.

Being market led does not mean satisfying every whim of every customer. There is some business we do not want, and there are some terms and conditions it would not be wise for us to accept. There are some minority requests for activities that it would be folly to build into the majority of our offerings.

We have to 'manage' our customer base to achieve our objectives. We have to be 'customer leading' as well as 'customer led'. We have to be 'market driving' as well as 'market driven'. This is a paradox. The art of marketing management is to steer a delicate course between the two conflicting factors (see Chapter 3) (Follow-up 1.2).

EXERCISES

1.1   Is marketing regarded as the means of generating profit in your organization? If not, is it because the marketing is wrong or because the perception of it is wrong? What needs to be done about it?

1.2   Are there some parts of your organization which appear to subscribe to the 'better mousetrap' philosophy? If so, what needs to be done about it?

1.3   Look at your company literature or other means of promotion such as web sites. Is the emphasis mainly on what your services *are*, or on what they will *do for the customer*? (See also Exercise 9.1.)

1.4   If marketing is too important to be left to the Marketing Department, are the other parts of your organization playing their part sufficiently in the marketing task? Where are the main areas of weakness? What needs to be done about them?

1.5   Think of some ways in which the profitability of your organization could be improved by:
increasing the size of the market
increasing profitability by *marketing* actions.

1.6   Is there a clear distinction in your organization between the roles of sales and marketing? Are both functions staffed by people of the appropriate calibre and experience?

1.7   Consider the five 'myths' in Section 1.12. Ask yourself, honestly and critically, if any of them seem to prevail in your company.

FOLLOW-UP

1.1   Is your organization likely to be recruiting any full-time marketing staff in the foreseeable future? If so, have those responsible for the recruitment prioritized the qualifications required, as in Section 1.9?

1.2   Observe the key decisions which are made in your part of the organization during the next three months. Do you think these decisions exhibit the right balance between market driving and being market driven?

2

WHY DO PEOPLE BUY?
VALUE AS PERCEIVED BY THE CUSTOMER

Key business issues   *Section*

O   Every buying decision is different. We must determine
    the influencing factors before we start selling   2.1

O   In some buying decisions, the product or service is not
    the main factor. In others, price does not dominate.
    'Place', promotion, packaging, perception, partnership
    or other factors may play a crucial role

O   The relative ranking of these factors can change with
    time

O   A major task of marketing is to bring price lower down
    the ranking in the buying decision   2.2

O   Customers do not simply buy a service alone but a whole
    bundle of attributes of which the service itself is only a
    part   2.3

O   Intangible factors such as image, credibility and perceived
    track record can dominate a buying decision

O   Peripheral aspects of a service may have more influence than the core features on which it is based

O   Personal relationships, such as consultancy and partnership, can override everything else

O   These three factors, particularly the last, can actually eliminate all competition

O   People buy *benefits* not *features*, but the conventional 'features and benefits analysis' has some serious limitations   2.4

O   Different people want different benefits from the same service. Benefits to some may be irrelevant to others and therefore dilute the message

O   A benefit to one person may actually be a dis-benefit to another

O   No single sales message can address all segments, but the cost of targeting need not be high if simple principles are followed   2.5

O   In a decision-making group, each member wants different benefits. The message needs to be targeted   2.6

O   We need to differentiate our service from that of the competition, and present it as unique (or at least special) in a way which is relevant to the customer   2.7

O   By following these principles, our sales and marketing activities can be given a sharper cutting edge   2.8

O   Branding gives 'personality' to a service, and defines the 'stable' from which it comes   2.9

O   Brand names may have a higher value than the tangible assets on the balance sheet. This can cause trauma in the boardroom in an acquisition situation

O   The marketing of intangible services requires some special attention, but it also has some special advantages   2.10

O Product companies might profitably add some services to
   their portfolio

O Marketing, properly understood and applied, is appropriate
   to virtually all fields of high technology service

2.1 THE MARKETING MIX

Before approaching a selling opportunity, we need to know the factors which
are going to influence the buyer. Every buying event is different, and generalizations are dangerous. To assume that buying decisions are always made on
the same basis will reduce the chance of achieving the sale.

Let us imagine that we are the manager of a large chemical engineering
process plant. We have our own internal technical resources, but from time
to time we come across problems which require outside expertise and we
feel that it might be possible to increase the profitability of our operation by
introducing improvements into the plant or the process. It is useful to analyse
the various factors that may influence our decision to bring in outside help.

The classical marketing mix, known as the '4 Ps', is made up of:

O   product
O   price
O   place
O   promotion.

To these we have added three more:

O   packaging or presentation
O   perception
O   partnership.

PRODUCT (OR SERVICE)

Which company can do the most appropriate job for us? Do they offer multi-
disciplinary tailor-made consultancy, or do they have a more basic standard-
ized approach? Do other aspects of the service they offer play an important
role in the decision?

PRICE

What fees do they charge? Are they the cheapest? Are we prepared to pay
more than the cheapest price if we believe it offers better value? Do we feel

that their fees are unreasonably high? Do their fees appear to be too low for the professional job we require?

PLACE

The 'place' in the marketing mix describes the means of contact between buyer and seller. For a high technology service, one of the most important factors may be speed of response. Is the service provider located near to us, or even in our country? Can we easily get hold of the person we want, or are we subjected to an impersonal coded answering system which seems to be designed more for the convenience of the supplier than the customer? Are they accessible in different time zones, 24 hours a day if necessary? Do any of these aspects of 'place' affect our buying decision?

PROMOTION

Promotion is any form of activity that surrounds the selling operation to increase the chance of a sale. In the case of our chemical engineering plant, how did we come to hear about that particular consultancy? Have we been influenced by visits, advertising, brochures, web sites or other material during our search? Has our opinion been affected by an article in a journal or a presentation at a conference?

PACKAGING OR PRESENTATION

For a tangible product – a dispensing package such as an aerosol, for example – the pack plays a key role in the use of the product and may cost more than the contents. This is obviously not the case when what we are selling is an intangible service.

However, it is useful to broaden the concept. In the case of a consultancy proposal, for example, the packaging might refer to the way in which the document is presented and structured, and the offer explained in a face-to-face presentation. Because it is more visible, the packaging may create more initial impact than the service itself. It has been said, 'you only get one chance to make a first impression'.

PERCEPTION

How do we perceive the service and the organization from which it comes? Is the brand name a strong influence on our buying decision? Do we believe that if a service comes from that particular supplier, it must be good, reliable and effective? Have we considered whether the supplying company will still exist in a few years' time, so that there will be no problem in obtaining continuing support and subsequent follow-up work? Do we perceive that the service offers good value for money?

Perception describes what goes on in the potential buyer's mind concerning the service and the company from which it originates. The image may be accurate, but it may also be false or be based on circumstances which no longer pertain. A politician, reporting on an opinion survey, stated that it was inaccurate, prejudiced and wrong. But if that's what people think, that is how they will vote! If our potential customers think that our service is the best or the worst in the world, or that we are expensive, slow to deliver, bad at providing support and so on, whether or not their opinion is justified, they will allow these factors to influence their buying decision just as much as the tangible components of the service itself.

PARTNERSHIP

One of the key elements in a successful business relationship is that buyer and seller regard themselves as partners. Both partners benefit, and it is a 'win-win' situation. One company has as its slogan 'think of us as part of you', i.e. 'call on us as readily as if we were an internal service department, but only pay us when you need us'.

There is nothing special about the seven factors listed, and the reader is encouraged to substitute others as appropriate. They do not have to begin with 'P'! The question to be considered is, 'what are the main factors, ranked in order, seen from the perspective of the potential buyer, which are likely to influence this particular buying decision?'

The point about the use of the word 'mix' is that the buying decision consists of a number of ingredients that can be combined in an infinite variety of ways. Just as a Christmas cake has a different mix of ingredients from a light sponge, so it is with buying decisions. It is a mistake to imagine that they are all the same, or to extrapolate from one situation to another.

The marketing mix can also change over a period of time. For example, in the early stages of a new service, the determining factors may be promotion, by which people are made aware of the service's existence, and perception in terms of the image with which the customer views us and the service we offer. Without either of these, the product and price may never be considered. Later in the life cycle, the task of creating perception has largely been achieved, and particular aspects of the service may assume first place (differentiation). Later still, when a number of competing services all do virtually the same thing, differentiation is hard to maintain and price may come to the top of the list (Exercise 2.1), (Follow-up 2.1).

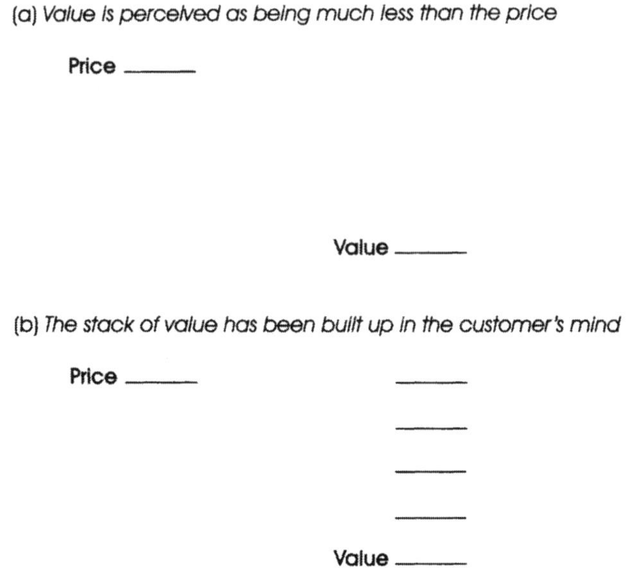

(a) *Value is perceived as being much less than the price*

Price _____

Value _____

(b) *The stack of value has been built up in the customer's mind*

Price _____   _____

                                _____

                                _____

                                _____

Value _____

Figure 2.1   Creating value in the mind of the customer

2.2   THE ROLE OF PRICE IN THE BUYING DECISION

Unless we have deliberately selected price as the differentiator, a major task of marketing is to bring price lower down the ranking in the marketing mix by pushing other factors upwards. A market leader may be able to justify a premium price by promoting the special attributes of the service and everything that goes with it. If we look at our own private and business purchases, we can easily find cases where we did not buy the cheapest service on the market.

One company bidding for a multimillion-pound tender came to the conclusion that there were ten factors likely to influence the award, of which price was number seven. It is not actually possible to rank the elements of the marketing mix as precisely as this, but the company is to be applauded for realizing that six factors might be more important than price in that particular situation.

In many cases, the most important factor is not the initial cost but the 'lifetime cost of ownership', i.e. the cost of continuing support, consumables, service, upgrades and so on. A service which is initially more expensive might actually be the better buy; the cheapest purchase might be the worst possible investment. If this is so, the argument will need to be convincingly presented.

If someone comes to our home and tries to sell us a new kitchen or double glazing, what happens if we ask the price early in the conversation? The request is ignored, because the salesperson is wanting to build a stack of value in our mind – to bring other elements in the marketing mix higher up the ranking order – before quoting a price. This can be illustrated diagrammatically (Figure 2.1).

In the first case (Figure 2.1a) the value is perceived as being much less than the price quoted – there is no correlation between the two in the mind of the customer. In the second case (Figure 2.1b) the stack of value has been built up; when the price is ultimately quoted, it is judged to be appropriate to the increased value which is now perceived.

This process requires all our skill, and is time-consuming. However, we have only two choices. Either we bring the perceived value up to the price we wish to achieve, or we bring the price down to the low value that is perceived. The effect on our profitability can be enormous, even to the extent of determining whether or not we stay in business. While many aspects of the double-glazing type of selling are totally inappropriate to high technology services, this price aspect is absolutely relevant.

We are not advocating overpricing or exploiting the customer, but we are pointing out that price is only one of many factors influencing the buying decision. In some buying decisions the lowest price normally wins. In many other cases people are prepared to pay more than the lowest price – sometimes considerably more – because they are convinced that the benefits offered to them justify the premium being asked.

The danger is to assume that all buying decisions are the same. This is a particular pitfall in organizations which are diversifying from an environment that was dominated by cost-plus or lowest price tenders into a free market situation; examples might be a research department whose original brief was to undertake programmes at a cost which was funded by the parent company, or a government laboratory which becomes an agency. People tend to extrapolate their thinking from the first scenario to the second. The simple fact is that much of the marketplace does not think in this way. By taking a market-based approach to pricing, it may be possible to increase prices in some cases and it may be necessary to reduce them in others. Most businesses have a mix of high volume/low margin and low volume/high margin sales; the objective is to maximize the total profit.

As we shall be discussing in Chapter 6, many companies overprice or underprice their services. This is because they have not clearly thought through the role of price in the buying decision.

2.3 CUSTOMERS BUY A WHOLE BUNDLE OF ATTRIBUTES

Many people make the mistake of thinking that what they are selling is restricted to the service itself. Clearly the service is a crucial component of the marketing mix – without it we have nothing to sell!

However, if we are honest, we have to admit that in many cases the service we are offering is not particularly different from the competition's. If this is so, we are not likely to succeed by trying to prove that our service is the best, and yet this is the instinctive reaction of most people faced with a competitive situation. In these circumstances, the service itself cancels out in the buying decision and other factors take precedence. The company that is better at marketing gets the order. This is very frustrating for people who have spent years developing the service, but we have to face commercial reality.

There are three main areas of influence that can override the service itself, namely:

O   intangible factors
O   peripheral factors
O   personal factors.

We shall consider these in turn.

INTANGIBLE FACTORS

Most people, such as the rational thinker, the scientist and the professional, tend to believe that customers buy what they can touch and see. Unfortunately, what customers cannot see is often more important than what they can. The fact that intangibles cannot be described with the same degree of precision does not mean that they are less important.

Intangible factors may include perception and image, peace of mind because a problem has been solved, or confidence because the matter has been entrusted to someone who is believed to be trustworthy. The service is not the ultimate objective of the purchase, but simply the vehicle by which the peace of mind or the confidence is achieved. If this is so, it is the peace of mind or the confidence that we should be promoting, not just the service itself.

In consultancy and the professions, the product is, in effect, the person providing the service. 'Personal chemistry' between the service provider and the client, for better or worse, may rank much more highly than an objective analysis of the skills being offered. Confidence is paramount, but this may be misplaced and it is certainly unquantifiable.

The need to market these intangible factors is important when we are trying to create a positive image, but it is even more so when we are trying to overcome a negative perception. A good image may take years to create, but it can be lost in one simple statement or episode and can take years to rebuild (Exercise 2.2).

PERIPHERAL FACTORS

Many buying decisions are made on the basis not of the service itself but side issues which are relatively trivial. In a high technology service, the differentiating factor may sometimes be technologically unsophisticated.

An American company for which I was working received a telephone call on Christmas Day asking to speak to a service engineer. The call was put through to the engineer's home where he was playing with his children. The caller then rang off. Early in the new year, we received an order for $100 000 for a piece of equipment. The buyer explained that he was the one who had telephoned on Christmas Day asking for a service engineer. Because we had one available, he bought the product. If we analyse that situation, we could say that the product we were selling obviously passed his buying criteria but so did the products from several rival companies. The overriding factor in his mind was the availability of service support at critical periods, because a problem with our product could virtually bring his operation to a halt. The highly sophisticated product, which had been developed over a period of years by PhD scientists, was not the final determining factor; it was the availability of a less qualified service engineer.

Knowing how customers value these peripheral aspects might help us to achieve a more effective use of our R&D budget. Some relatively minor amendments, created at a relatively low cost, might make the service much more acceptable (Exercise 2.3).

PERSONAL FACTORS

Personal factors may include the way in which a salesperson relates to a potential buyer, but there is much more to it than this.

The goal we are trying to achieve by promoting personal factors is a marketeer's dream – to eliminate all competition, not only from the immediate buying decision, but also for an indefinite period thereafter.

I once received a telephone call from a manager in one of the UK's leading companies. He said, 'Hello Colin. Dave here from ... – how's your diary?' We looked at convenient dates and arranged a seminar. There was no written invitation to submit a competitive tender to be judged on a certain date by a panel – the only issue was when I was available. On that occasion at least, there was no competition. Obviously we have to earn such a position in the first place and we have to nurture our customers to maintain that privileged status but, as long as we can do so, there is a high probability that we shall continue to receive their business.

There are various ways in which this personal relationship can manifest itself. These are worth deliberately cultivating.

First, we might be seen as an expert to whom the buyers turn when they have problems. They want the difficulty to be removed so that they can get on with what they are really there to do. If we can quickly get to the heart of their apparently insoluble problem, we become an indispensable ally. An important step in the process may be to help them realize that their problem is not what they thought; if we help people to identify their real problems, we may not always be guaranteed the business of solving them but we are surely in 'pole position'.

Second, buyers might regard us as the consultant to whom they turn for advice, quite outside the context of a particular consultancy project. They need our objective view and broader experience. One way of nurturing such a relationship may be to create occasional opportunities for small amounts of free consultancy. The aim is to 'keep the door open' and to maintain a positive profile even when no immediate business is apparently being offered.

Third, we can be seen as a business partner. This comes about when we have a growing relationship with our client, so that we become involved in the longer-term and often confidential strategic decisions that our client is facing. Our knowledge and our privileged position form an entry barrier to competitors, and the partnership strengthens as we become more closely involved with the client's business.

The aim of these three forms of relationship is to establish a position where we are the first and, hopefully, the only person to whom the buyer refers when the need arises. The ultimate aim is to be regarded as if we were part of the buyers' operation, so that they turn to us as naturally as they would make an internal telephone call to a colleague (Exercise 2.4).

2.4 THE FEATURES AND BENEFITS TECHNIQUE AND ITS LIMITATIONS

Almost every book and every course on marketing deals with features and benefits. However, many do the technique a disservice by presenting it in an inadequate manner. We shall present the argument, discuss its limitations and then go on to show how, when properly applied, a features and benefits analysis can be one of the most powerful tools in marketing.

Let us illustrate the technique by referring to high technology problem-solving consultancy. Features might include:

- world-class scientific skills
- multi-disciplinary expertise
- intimate knowledge of the client's business
- research backup to the technical service offered
- 24-hours a day availability
- proven confidentiality, and so on.

The argument is that all these things are *features*. People do not buy features, they buy benefits. Features are *what the supplier puts into* the product or service; benefits are *what the customer gets out* of it.

A useful link between features and benefits is the phrase '*which means that*'. So, in our problem-solving consultancy example:

O   It has world class skills, *which means that* the client receives the best solution.

O   It offers multi-disciplinary expertise, *which means that* the client has access to a 'one-stop shop'.

O   It has intimate knowledge of the client's business, *which means that* the client does not have to pay for the learning curve, and the work is completed more quickly.

O   It has research backup, *which means that* the client can benefit from the latest technology.

O   It has 24-hours a day availability, *which means that* emergencies can be dealt with quickly.

O   It has proven confidentiality, *which means that* the consultant can be brought into discussions of strategic importance and can make a greater contribution to the client's business, and so on.

Note that some of these benefits are intangible and unquantifiable, but they may actually weigh more heavily with the client than some of the tangible benefits.

What we are doing is to move away from the service itself to the issues which are of concern to the potential customers. They do not want consultancy – they want what it will do for them.

This involves detaching our thinking from the service we are selling and taking the trouble to find out how the buyer thinks. For some people, such as high technologists, professionals or inventors, this is quite a difficult task. They can be so absorbed in the world of technology, procedures, specifications, regulations and so on, that they find it hard to think beyond these to the concerns of the customer.

The features and benefits technique is normally presented as part of the selling process. Salespeople are encouraged to sell the benefits rather than the features. However, the technique is much more than a sales aid, which is why it is included at this stage in the book. The right time to start thinking about features and benefits is at the concept stage in the design of a new service and the development of its marketing strategy. Customer benefits can be often built into the service at this stage much more easily and cheaply than later. They may not involve the highest technological aspects of the service, but they can give it a cutting edge in the marketplace.

Benefits ultimately provide value. The more value we can create for the customer, the more profit we are likely to generate for ourselves. (In most cases the value and profit are expressible in financial terms, but this is not necessarily so; the technique can equally be applied to a non-financial situation such as safety.)

To make best use of the features and benefits technique, we should take the analysis as far as we can in the customer's direction. For example, we have said that multi-disciplinary expertise is a feature, and a one-stop shop is a benefit. However, it would be better to say that a one-stop shop is actually a more refined description of a feature, rather than an ultimate benefit. We can therefore transfer it back to the 'Feature' column, and continue the process:

Feature	Benefit
	which means
multi-disciplinary expertise	one-stop shop
one-stop shop	total solution from one supplier
total solution from one supplier	quickest solution and least disruption
quickest solution and least disruption	better plant utilization
better plant utilization	more profit

In high technology business, 'more profit' is usually the ultimate objective. In presenting our case, these financial and business benefits should be promoted rather than the technological features from which they derive.

One way of arriving at a benefit is to think what the dis-benefit would be if that feature were absent. Thus, in this example, if the 'one-stop shop' feature were absent, the client would have to search for a number of consultants each able to solve part of the problem. This would take time, cost money and might even lead to disputes between different contributors to the same project. Offering a one-stop shop avoids all these problems, and the client might well be prepared to pay extra for it.

We do not have to use the word 'benefits' in our literature or our speaking (but why not?), and it is not always appropriate to show a double column listing the features and benefits. The important factor is to use the language of the customers, refer to matters which are important to them, and concentrate on the benefits which derive from using our service rather than on the service itself.

We now go on to show that, if used superficially, the technique can fail completely. The reason is very simple. We cannot make an automatic link between features and benefits – it depends on the buyer and on the situation in which the service is to be used.

First, *different customers want different benefits from the same service,* sometimes even from the same feature. The fact that we have a major research establishment is of no benefit to someone whose plant is leaking and urgently requires a 'quick fix'. To others, our research backup is what they most need.

Should we then list every conceivable benefit we can imagine, in the hope that the customer will be attracted by at least some of them? The problem with this approach is that the message is diluted. The buyer may be distracted by irrelevant claims, and fail to reach the very one which would have been convincing.

Second, and even more seriously, *something which is a benefit to one person may actually be a dis-benefit to someone else.* Consider, for example, a situation where we are the largest supplier in the world of certain skills. Some potential customers may actually perceive this to be a disadvantage. They believe (perhaps quite wrongly, but that is not the point) that we are slow and bureaucratic, expensive and over-thorough. What they want is urgent action at an economical price. Our status and past history are of no relevance to them and may instead count against us.

Benefits depend critically upon the presentation. Obviously they have to be based on fact, but we could produce two messages about the same service which portrayed entirely different benefits; when we come to consider market segmentation in Chapter 4, we shall see that this is precisely the approach which is recommended.

We are beginning to make the point, which we shall develop throughout the book, that successful marketing is not like firing a blunderbuss in all directions but is more like aiming a sniper's rifle at different targets in succession with different ammunition in each case.

We are not offering an easy life to marketing staff. It is much simpler to use a standard sales pitch or produce a standard brochure than to tailor our message to the situation. However, the more trouble we take to marshal the arguments, present the benefits and, especially, target them to the needs of a particular reader or hearer, the more successful we are likely to be (Exercise 2.5).

2.5 DIFFERENT MARKET SEGMENTS WANT DIFFERENT BENEFITS

Let us suppose that we are selling a laboratory testing service. We might list all the features and their corresponding benefits, prove their validity in terms of performance tests and so on, and illustrate them lavishly with pictures.

The problem is that people want laboratory testing for very different reasons. For some, the main benefit they seek is speed and reliability. They want to use an outside resource to provide routine control for their own operation, and to do so at an economical price. Others want outside help because they need access to equipment so sophisticated that it would not make sense to buy it for themselves.

It is impossible to construct a single message that will meet the needs of both the basic and sophisticated segments. Having broken the marketplace into segments (which can be done in a variety of ways – see Chapter 4) we need to try to determine what will motivate each segment, and present the benefits accordingly. This targeting of benefits obviously increases the complexity of the marketing task, but we can group customers together so that we do not have to produce one brochure per customer! In Section 4.5 we suggest a way in which a family of brochures can be produced without incurring excessive cost and effort; the same is true of advertising, web sites, sales presentations and most forms of marketing communication.

In the course of my consultancy work, I frequently see literature that tries to cover so many targets it does not hit any of them effectively. This is not a good way of spending marketing budgets (Exercise 2.6).

2.6   EACH MEMBER OF A DECISION-MAKING GROUP WANTS DIFFERENT BENEFITS

It is a mistake to think that everyone who influences the buying decision comes from the same background as the seller. Technical people tend to think quite wrongly that the main message is always a technical one. Some experts seem to assume that their clients understand and are interested in the jargon they use in their particular profession. These attitudes create a barrier between buyer and seller.

In many buying situations, the decision is made not by one person but by a number of people who constitute what is known as the decision-making group (DMG) or decision-making unit (DMU). The DMG will be represented by various departments, functions and job titles – technical, financial, operations, budget holders, senior management, quality and so on.

As with the market segments, the task is to identify the individuals and departments involved in the buying decision, decide which benefits will motivate each of them and present the message accordingly. In a large contract involving a long drawn out negotiation, this might involve separate presentations to different members of the DMG.

In other cases this approach is not practicable, and one brochure, letter, proposal or tender submission will have to serve all purposes. We need to find a means of presenting the benefits differentially, by having clearly identified sections covering the issues relevant to the different members of the DMG. We cannot assume that anyone will read the whole document – indeed, we do not necessarily want them to. We want each of them to be motivated by the particular points which are most relevant to them. The fact that the message is hidden in there somewhere is not good enough – each busy reader must be able to find the relevant part quickly.

In a face-to-face presentation where all members of the DMG are present at the same time, we must again have material available for each of them and present it in the right way at the right time.

The main issue with a DMG is that, in principle, any one of its members can effectively veto the buying decision. To convince all but one may result in failure.

Apart from the formal DMG, there are usually some people who have no authority to sign a contract but who can exercise a strong positive or negative influence on the decision. Users are a typical example. If they do not like what we are offering, they will make their views known and thus effectively have the power of veto. If they like it, they will say so. Nurturing the interests of these influencers may be an essential part of the marketing process, and it may be useful to create opportunities to talk with them.

Another form of segmentation familiar to many high technologists is the split between contractors and subcontractors. Each has an entirely different view of the project, and we need to market differently to them. If we are subcontractors, we need to market ourselves to both the main contractor and the ultimate client but, again, the way we do so and the message we communicate may be very different in the two cases.

The make or break parts of a negotiation may be carried out at various levels of seniority within the two companies concerned. The better we are at marshalling all our resources and ensuring that they present an appropriate message to each member of the DMG, the greater will be our chance of success. The award of a large contract is a multi-level multi-disciplinary process for both buyer and seller; winning it is a job for marketing and management, not just for salespeople (Exercise 2.7).

2.7   THE UNIQUE SELLING PROPOSITION

The unique selling proposition – the 'USP', sometimes known as the unique selling point or the sustainable competitive advantage – is both elusive and crucial. It asks, 'what is unique about our service?'

The philosophy is very simple. If we have correctly identified marketplace needs and if we uniquely meet those needs, the business will be ours. Unfortunately it is not usually as easy as this!

Semantically speaking, a service either is unique or it is not. It cannot be fairly unique, very unique, nearly unique or more unique than some other service. In the realities of business, however, very little is truly unique. What we are seeking to do is to differentiate our service from that of the competition. What is special about what we have to offer? Can we promote this USP as a reason for buying? Can it justify a premium price?

We must recognize that it is a unique *selling* proposition. Uniqueness that is not valued by the purchaser is not a USP. We may have the only piece of equipment in the world that performs a certain task; that ability is unique, but it is not a selling proposition if that particular buyer does not want the task to be performed.

Some attributes that we regard as unique might actually have a negative connotation in the minds of customers. Claiming that we are the oldest-established company in our field might give us a good feeling, but to a potential customer it might suggest that we are out of date in our technology compared with newcomers.

It is very difficult to claim differentiation as a general practitioner (except, perhaps, on the grounds of 'one-stop shopping'). If we have a wide-ranging expertise, we may need to present ourselves as offering multiple specialisms and marshal an argument for meaningful differentiation in each.

As we have discussed in Section 2.3, differentiation can be based on intangible, peripheral and personal factors as well as on the service itself. Our service may be faster or more effective, but it may also make the buyer feel better, people may like the way we manage a project or it may be sold in a more professional way than that of the competition.

The USP is, in a sense, the ultimate form of benefit. It needs to be targeted in the same way to different market segments and members of the DMG. This form of differentiation does not require a highly scientific or intellectual approach, and it should not be despised on that account; it may make all the difference as to whether our service succeeds or not (Exercise 2.8).

2.8 THE SHARPER CUTTING EDGE

Figure 2.2 summarizes the issues discussed in the last four sections. It shows the way in which we can give our marketing efforts a 'sharper cutting edge'.

Marketing is about probabilities. There are no techniques that automatically give results, but thoughtful application of this approach will increase the probability of success.

The basis of the sharper cutting edge is, of course, features. Without features we have nothing to sell. They may be represented by the skill-sets we are able to offer, the equipment we use or some particular computer programs we have developed. Features are the foundation of our marketing efforts. However, used alone as they often are, they have a blunt cutting edge.

Moving up Figure 2.2, we come to 'benefits'. As we have said, these need to be targeted in two ways – to market segments and to members of the decision-making group. The peak of Figure 2.2, the sharpest cutting edge, is the unique selling proposition.

The more we can disentangle ourselves from the features, the better we can target our benefits to the needs of each particular customer. The better we can differentiate our offering by some form of uniqueness, the more effec-

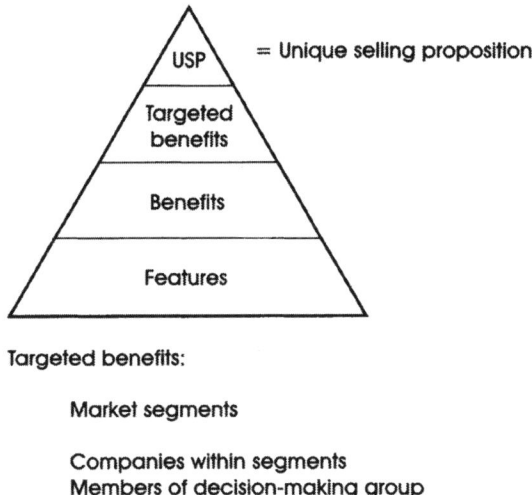

Targeted benefits:

    Market segments

    Companies within segments
    Members of decision-making group

Figure 2.2 The sharper cutting edge

tive will be our marketing (Follow-up 2.2).

I am something of a rebel concerning some aspects of corporate identity programmes. The theory is that everything we do should reinforce the same identity. I used to have a colleague who said, 'I have my principles, but if you don't like them I have others!' We may need to present a different identity to different market segments. For example, when it is beneficial to present ourselves as the largest international organization in the field, we should do so; in other cases, it might be much more appropriate to show that we are fast-on-the-feet operators who can act independently without bureaucracy or other constraints.

2.9 BRANDING

If we heard that a consultant came from a particular consultancy, there are certain assumptions we would immediately make. These might concern the qualifications, the experience, the professionalism of the person concerned and the fee charged for his or her services.

This is because we associate certain qualities with the 'stable' from which a service or a service provider comes. Marketing efforts are directed not so much at the service itself as in developing the 'personality' of the brand. This personality may be embodied in tangible factors such as 'outstanding performance' or in intangible factors such as 'makes you feel confident'.

A key element of successful branding is that our service is positioned differently in the mind of potential customers from the services offered by competitors. The differentiating factor must be unique, simple to grasp and not easily confused with competitive claims. The aim is that our service is perceived as dominating that particular marketplace.

The supplier engages in advertising and other marketing activities in order to build the personality and positioning of the brand. Branding is usually embodied in the name of the company, but it may sometimes be in the name of the service. Assuming that the claims are consistent with perceived performance, the message is cumulatively reinforced by these marketing efforts.

It may be necessary over a period of time to modify the personality or move the positioning slightly, but this should normally build upon rather than conflict with the earlier claims. Exceptions would be when a disaster has occurred, or when the whole basis of the organization has changed, in which case the past mould must be broken and a new identity established.

A good brand image can be one of the greatest assets a company possesses. This is very relevant to an acquisition situation. If we are buying a company, the prime objective is not normally to acquire their facilities but to buy their market share, their customer base and their reputation, which are parts of their brand equity. One of the biggest mistakes an acquiring company can make is to change the name of the company. At a stroke, they may be destroying the very asset they have acquired. The value of the brand causes problems for directors or accountants who complain that the tangible assets they are buying are worth less than the price they are being asked to pay, but this reflects reality. We can build new facilities more quickly than we can build a brand identity, particularly in a world-wide marketplace. The premium we have to pay is technically described as 'goodwill', but it encompasses far more than this rather vague accountancy term suggests.

Market-led companies often develop a powerful brand name in one particular field and then extend its use into other quite different service areas. The common thread may be that they are appealing to the same socio-economic market segment. The fact that the new services have nothing to do with the original service is irrelevant. While the process may be gradual over a period of time, the ultimate result is a significant shift in the balance of the company's operation. Other companies may need to follow this example (Exercise 2.9).

2.10 SOME DIFFERENCES BETWEEN INTANGIBLE SERVICES AND TANGIBLE PRODUCTS

It has been made clear from the start of this book that the basic principles of marketing apply equally to tangible hardware products and to intangible services. The reader has been encouraged to regard a service as a 'product'.

Nevertheless there are certain problems and tactics which are specifically associated with the marketing of services. There are also some advantages in marketing a service, and hardware companies might find an appropriate range of services a useful adjunct to their product portfolio.

If we are marketing a physical product, such as a video recorder or a car, we can demonstrate it, give the potential customer an opportunity to try it out and, if appropriate, lend it for a period of time. With activities such as consultancy, professional services, dedicated software and so on, none of these methods of convincing the customer is available. The 'product' does not even exist until we have performed the service. How do we overcome this? We shall consider several aspects.

INTANGIBILITY

The main consideration concerning a service is that it is intangible. From its Latin derivation 'intangible' means 'cannot be touched', but the problem is much deeper than this. The service cannot easily be judged or compared with alternative offerings. Its performance cannot be measured in advance. It cannot offer the comfort that a buyer experiences if a product can be seen and tried out before purchase.

Somehow we have to generate a package of evidence which acts in the same way as the tangible aspects of a piece of hardware. The best way of doing this will depend upon the nature of the business, and confidentiality may have to be considered.

Lists of current and former clients may be compiled, but it may not be appropriate to tout these around.

Clients may be willing to write testimonial letters, and it is perfectly reasonable to ask them to do so. We may have to promise some limitations on their use. A reasonable approach, which clients rarely reject, is to ask permission to show the letters to non-competing potential clients, but to undertake not to use them for advertising or to send out copies unsolicited.

It may be possible to describe a number of case studies of relevant past projects; we may or may not be able to use the name of the clients concerned.

It is very helpful if we can quote reference sites where a potential customer can contact a previous purchaser and, possibly, even visit the site in order to see the result of the service at work. Some customers are very happy, within reason, to show and talk about what they have; in other cases, confidentiality may preclude this. The privilege must not be abused, and reference sites should be used very selectively otherwise the goodwill may be destroyed.

Establishing a relevant track record is crucial to the marketing of intangible services. This aspect of marketing is developed in Section 4.8.

UNCERTAINTY ABOUT THE REPUTATION OF THE SERVICE PROVIDER

Compared with fast-moving products, the number of customers and the frequency of purchase may be so low that a high profile in the marketplace is very difficult to establish. The potential client may not even have heard of the service provider, and finds it difficult to form a reliable and objective view of his or her calibre.

Effective ways of raising our profile include writing 'learned papers' in the appropriate journals, submitting interesting articles to the press, presenting at conferences, running educational seminars and so on. Although these activities are time-consuming, they can be at least as effective as other ways of spending the marketing budget. To avoid undermining their authority and credibility, they should be deliberately divorced from a selling environment. We can, however, make use of them after the event, for example, by sending out reprints of articles to a database of existing and potential clients.

MATCHING DEMANDS WITH RESOURCES

Any business finds matching demands with resources to be difficult, as customers do not place orders at a uniform rate exactly when we would like them to. With a physical product, we can build for stock if demand is low, or offer delayed delivery times if demand is high, but we cannot do this with a service. A consultant cannot be in two places at once. An empty seat on a seminar cannot be occupied by two people next time. Specialist software cannot be written 'for stock'. The operation has to be performed with the resources available when the client wants it. One director of a large engineering consultancy firm lamented, 'We always have 50 too many or 50 too few consultants'.

While it may never be possible to remove this problem completely, a helpful step might be to develop a portfolio of services, some large and some small. In general, the larger the project the longer the lead time involved in negotiating for it. Conversely, it may be possible to obtain some small consultancy contracts relatively rapidly to fill in the 'profit gaps'. Although the value of these projects may not be large, the percentage margin may be much greater than in a large contract which generates intense competition as every supplier desperately tries to get the business, and they may represent a useful stepping-stone towards larger contracts later.

PRICING

With a tangible product, such as a piece of domestic or industrial equipment, we can form a reasonable idea of the price – within, say, 20 per cent – by looking at it and comparing it with competing products. With an intangible service, unless we have had considerable prior experience in the particular area, we might be uncertain of the price by a very large percentage, possibly a factor of two, three or even more (see Chapter 6). Does the non-specialist have any idea what is the 'right' price to pay for an expert witness, for registering a patent in Germany, for undertaking a safety audit or for help with writing a business plan?

The absence of an obvious price reference for a service is not necessarily a disadvantage. We should not take advantage of the client's ignorance, but we do have the opportunity to create real perceived value and to price accordingly.

RAPID RESPONSE, FLEXIBILITY AND LOW RISK

Those who have worked with physical products are all too aware of the long lead times involved in the conception and development of new products. When the specification is finally agreed, we are faced with a long period of agony before we know whether or not the product is going to be commercially successful. There is a high risk of and a high penalty for failure.

Intangible services gain in all of these respects. A 'new' product can be developed in a matter of days, amendments can be made at the touch of a keyboard, and the time from conception to launch can be a matter of weeks. Those involved with the marketing of services must exploit these enviable advantages to the full.

A profitable route for business development for companies in the tangible product field may be to add a range of services to the portfolio. For example, those who have developed expertise in modern manufacturing technologies could offer consultancy and training in these technologies (preferably to non-competing companies!). In this way they might generate gross and net margins that make their manufacturing operation seem rather unexciting by comparison.

To summarize, the principles of marketing, properly applied, are absolutely relevant to consultancy, professional and other services. In one sense they may be even more necessary, because there is no tangible product which can draw attention to itself.

EXERCISES

2.1   Choose two different services sold by your company, and rank in order the three elements of the marketing mix which you think are most important in the purchasing decision. Note the ways in which the ranking differs in the two cases. How would you exploit the first on the list? (For example, if you think that perception is top of the list, what specific actions could you take to improve perception?)

2.2   Make a list of the intangible factors that influence the buying decision for your services. How could you make use of these?

2.3   Make a list of the peripheral factors that influence the buying decision for your services. How could you make use of these?

2.4   Make a list of the personal factors that influence the buying decision for your services. How could you make use of these?

2.5   Choose one of the company's services, and list the features and the corresponding benefits, using the link 'which means that ...'. Ask yourself if some of the benefits you have listed could be described as more refined features, and apply the process again to take them as far as possible in the client's direction.

2.6   List the market segments into which one of your services might be sold. How would you market differently to each segment? (Also see Exercises 4.2 to 4.5.)

2.7   Choose a sales opportunity and list the members of the decision-making group. Note the ways in which you would market differently to each member of the group. List the people who cannot make a purchasing decision but who could exercise a negative or positive influence.

2.8   Sit down with a colleague, or a number of pairs of colleagues, and explain the concept of the unique selling proposition. Ask one member of each pair to try to convince the other that there is a USP in what they are selling in their business life; the other should do everything possible to destroy the claim. Reverse roles. Ask each person to say whether, when they were on the receiving end, they accept that their partner succeeded in demonstrating a USP. The percentage success rate can be very revealing.

2.9   What are the main components of the brand equity of your service? Are you exploiting the personality and positioning of the brands to full advantage? What further steps could you take? Could you use the brand as a vehicle for other activities?

2.10   If you currently market any services, what steps can you take to make the 'product' appear more tangible so that a potential client would be in a better position to assess its value? Are you sure that the services are being priced correctly? (Also see Exercises 6.7 and 6.8.) Think of additional services that you could market to complement your present portfolio.

FOLLOW-UP

2.1   For the next three months, keep a record of typical buying decisions which you make in your business and private life (including those where you decided not to buy). Note the factors that most strongly influenced the decision. If you did not buy, was there anything the seller could have done which would have made a difference?

2.2   For the next three months, keep a record of selling transactions in which you participate. Note against each where you think they rank in Figure 2.2.

3

WHAT IS INVOLVED IN BECOMING MARKET LED?
SETTING OUR SIGHTS HIGH

Key business issues	Section
○ A local, technical, sales or finance led orientation is a very inadequate substitute for being market-led	3.1
○ Becoming market-led involves significant culture changes	3.2
○ We can be reactive and complacent, beginning to worry, limited by attitudes and systems, unwilling to make the commitment, panicking; or we can be truly market led	
○ A strong marketing function interacting synergistically with other equally strong functions can transform a company and be a powerhouse for business development	3.3
○ Those who have experienced this synergy would never settle for anything less; those who have not experienced it have never really understood marketing	
○ We should seek to maximize the profitability of existing resources before acquiring more resources	3.4
○ Targets and resources need to be compatible	3.5

3.1   THE DEVELOPMENT OF THE MARKET-LED PHILOSOPHY

It is helpful to consider how modern marketing thinking has evolved. This will enable us to see where our company stands, and to decide whether any changes need to be made.

Deeply rooted attitude and cultural issues may be found at the highest level. Middle managers cannot make progress if they are inhibited by the conservative attitudes of their senior managers.

A LOCAL OPERATION

The simplest case is exemplified by a village bakery. It makes bread, cakes etc. and people buy them. Customers come to the bakery – it does not have to go out to the customers. Customers buy every day, and it is obvious if one of them stops buying. They talk face to face, and there is no need for all the complexity of modern marketing communications – literature, advertising, web sites, exhibitions, public relations and so on.

Innovation for our locally oriented bakery is simple, rapid and virtually risk free. They make a few new cakes or loaves and display them on the counter. Within a few hours they know the level of the initial interest, and within a week or two they will have some idea of the potential for repeat sales. This contrasts with the intensely complicated process of new product development today, often involving massive investment and high risk.

With these enormous advantages, a local emphasis would appear to be very attractive. However, it also has problems, which can be serious enough to kill a business.

First, locally led marketeers are at the mercy of external forces which are outside their control. For the baker, the population of the village may be declining, growing old or moving away to the city. For us, there may be changes in legislation, standards, technology, availability of finance and other factors which are outside our control but which can materially affect our business prospects.

Second, a new competitor may set up in opposition, so that the sales potential is reduced.

The problem of a local emphasis is well known to professionals who historically served their local clients who generally came to them and kept them busy and financially rewarded. Now, many of these professionals are compelled to market their services proactively and take initiatives to acquire new clients.

Another variant of a local emphasis is seen in operations that sell only in the UK. While this may offer enough potential for some companies, many are finding that they have to target at least part of the international market.

A TECHNICALLY LED OPERATION

In areas where a high degree of expertise and experience is involved, there is a great tendency for people to be inward looking. We have discussed the fact that the expertise and experience dominate our thinking to such an extent that it becomes very difficult to give due weight to the needs of the external marketplace. It is a classic example of 'in-to-out' thinking.

With increasing sophistication of scientific expertise, a whole set of management skills has been built round the research and development operation. This can involve a large investment in people and equipment, with the result that the internal operation tends to become the focus of senior management attention. The instinct is to concentrate on 'what we do' rather than on 'what they want'. Simply increasing the internal capability without establishing a market demand is a pointless and very costly exercise. An inward-looking emphasis on meeting our own needs is destined to fail if these do not coincide with the needs of the external market.

In these cases, the main emphasis of brochures, advertisements, web pages, tender submissions, proposals and other forms of communication is traditionally on telling the readers about ourselves, our resources and our past history. There is little identification with the customers' needs, and readers find it difficult to relate the message to their own business.

These inward-looking attitudes can actually be encouraged by our education and training if these are not broadly based. Technical and professional education is only just beginning to put vocational skills into the wider business context.

Whether we like it or not, the simple fact is that many contracts go to suppliers who take the most trouble to identify with the customer, even if an objective analysis might show that their skills were in some way 'inferior'.

A SALES-LED OPERATION

If we have a warehouse full of products, or a team of consultants without fee-earning projects, the instinct is to go out and sell. The battle-cry is 'sell, sell, sell'. What is wrong with that? Surely it is the right thing to do?

The problem with a sales orientation is that it assumes that we are selling the right product to the right people at the right price with the right claims, which may not be the case. I am reminded of the politician who, knowing that a forthcoming speech contained a weak point, wrote in the margin 'argument weak – shout louder!'

It is significant that the subject of selling is not discussed in detail until Chapter 8 of this book. The broader marketing and business issues that we are now considering are an essential prerequisite to a successful selling operation.

A sales-led environment can be very frustrating for managers with a mature understanding of marketing. Harder selling and more aggressive negotiations are not the answer to every problem. Taking time to stop and think may make those selling and negotiating efforts much more productive.

A FINANCE-LED OPERATION

We have already stressed that marketing is a very financially oriented discipline. One of the paramount objectives of most operations is to remain financially sound – to make profit, to control costs, to increase the value of the company's equity and so on.

These facts have important implications for the way business decisions are made. They will indicate the influence which financial management should have on strategic and tactical decisions.

Nevertheless, these financial objectives cannot be met if they are not consistent with marketing objectives. Financial tools and models are a delusion if they are not based on marketplace realities. The customer does not owe us a living. We can control our costs down to a penny, and still fail completely because we have not been able to generate income from the marketplace.

This is not to suggest that marketing departments should be free to violate budgetary control or to be casual about the way in which they spend company money. If anything, marketing managers should be *more* conscientious about financial matters than the financial departments themselves; in most organizations, marketing is making decisions about spending the money, while the financial departments are merely recording and commenting on where it is being spent. Marketing management needs to be intimately concerned with crucial issues such as investment, pricing, profit, cash flow and risk – these are central to the philosophy of marketing.

A MARKET-LED OPERATION

A market-led operation may also be referred to as being marketing led, consumer led or customer led. The key emphasis is on the needs of the marketplace. The first questions we ask are 'what are those needs?' and 'which of them are we going to meet?' We then ask 'what are the implications for our organization?' Until we have answered these questions, we are not in a position to start doing research or development, or to build up expertise or to start selling. We have to correlate the needs of the marketplace with the priority and the resources that we are willing to commit to meeting those needs.

Being market led is really the subject of this whole book. It has far-reaching implications for our organization, attitudes, culture, salary scales, management style, recruitment – indeed almost every aspect of our business. Ultimately it may be the biggest single factor affecting our profitability (Exercise 3.1).

While we should be market led, we should at the same time in some senses be market-leading or market-driving. It has been said that 'marketing is making the future happen'. Although we depend totally on our customers to enable us to do that, we cannot leave the initiatives to them. Considering the implications of this paradox for our particular operation can be very enlightening, and can help us to achieve the right balance for our attitudes and culture (Follow-up 3.1).

A research scientist in one of the UK's leading high technology companies once said to me, 'the problem is that I am working to a ten-year time-scale but I have to sell my ideas to accountants who have a one-year time-scale'. The dilemma is perfectly understandable and is, ultimately, a failure of senior management. The problem is that there is often no one on the marketing or the financial side working to the same time-scale as the long-term scientific staff. Why not? Should there be? How can the right business decisions be made in their absence? The fact that the market does not know what it wants in ten years' time does not absolve us from trying to get into customers' minds to discern what would be of value to them if we could provide it.

There is a myth that technical people create wealth and marketing people dissipate it! This view is based on an unbalanced view of business, but unfortunately it is common.

3.2 CULTURE CHANGES INVOLVED IN BECOMING MARKET LED

There are some common stages that companies may go through in the process of becoming market led. The notes that follow are based on my observation of a large number of companies; no reference to a particular organization is intended.

(A) REACTIVE AND COMPLACENT

Buyers come to the seller, and the seller's resources are completely utilized – they do not need any more business. This is fine as long as the demand is sustained and the competitors do nothing to undermine the position. Unfortunately, with such a self-satisfied attitude, many operations have declined into continuing struggle for survival and some have failed altogether.

This reactive approach has been common in some professional practices, including those in the high technology field. They have historically built their business through reputation and personal contacts. They have been brought up to feel – and the code of conduct of some professions actually encourages this feeling – that soliciting business conflicts with the dignity of their profession. The desire to avoid professionals making great efforts to 'poach' each other's clients is understandable, but these fears about marketing are based on a misconception. Some highly ethical and professional operations have to solicit all of their business, but they would insist that they do so in an entirely professional and dignified manner. We would not market management consultancy, for example, in the same way as double-glazing or time-share.

A similar circumstance arises with the privatization of nationalized industries or the move towards agency status of former government and ministry operations. It is also occurring in the normal industrial world where, for example, a scientific research unit is being told that it can no longer rely for all of its work on the parent company; the parent is free to use external service providers, and the research unit is required to generate new business in the competitive free market. In cases such as this, attitudes towards marketing are being compelled to change; it is no longer sufficient to sit back and wait reactively for internal clients to approach them.

As we have said (Section 1.10), much of marketing is aimed at increasing the total size of the market. If a large number of people are not taking out patents, considering environmental issues, upgrading their plant and processes and so on, what is wrong with promoting the idea that they should? Those who assume that marketing only involves fighting our competitors for known opportunities reveal an inadequate view of marketing. If the total market size can be increased, there is more for everyone – or rather for those who have the vision and initiative to create this new business.

Some people say 'marketing does not work in our business'. Inadequate, inappropriate or inept marketing does not work in any business. We have to decide which particular form of marketing is right for us.

(B) REACTIVE AND BEGINNING TO WORRY

People in this group are becoming worried by the fact that demand for their service has dropped off. They tend to blame outside forces such as the economy, the government or legislation.

Rather than investing in marketing, the instinctive reaction of many people in this category is to cut costs. Of course there are times when cost cutting is appropriate, but there is a danger that we merely slow the downward spiral without solving the real problem. Paradoxically, the better solution might be to spend *more* money, wisely and in acceptable quantities, to break the vicious spiral by *creating* new business opportunities.

(C) WANTING TO BE PROACTIVE BUT LIMITED BY ATTITUDES AND SYSTEMS

There is a real conflict in the minds of people in this group. They want to expand their business and know that this will involve some extra effort and expenditure, but internal rules and attitudes prevent them from doing so. We have already referred (Section 1.1) to the objection that 'marketing is a debit against profit'.

Companies in this category usually have no staff dedicated full time to marketing activities. This is typical of consultancies offering management, engineering, design, scientific research and other services. Virtually every member of the operation above a certain level is there to generate fees. Someone once asked me 'are you really suggesting that I should take people off fee-earning in order to do marketing?' as if I were crazy. Someone has to do it.

Managers with negative attitudes have a lot to answer for in these situations, as the following examples show.

Some companies, dominated by 'bean-counters', insist that any time or expenditure aimed at generating new business, such as visiting a prospective client or writing a brochure for a new activity, is booked against an existing project. This is pathetic! This exercise is not only pointless but highly deceptive, and it is they themselves who are being deceived.

Statements are made such as 'we are not allowed to cross-subsidize different activities'. This is a misunderstanding of how businesses develop. The marketplace and the competition are not interested in how we do our accounting. While we are playing internal games, the competitors are laughing all the way to their order books.

Some companies insist that 'all contracts must make a profit, even the first in a new area'. At first sight this seems to be a very reasonable philosophy, but it represents a serious trap for those trying to diversify from their traditional areas. A company might see opportunities for supplementing its defence business with the equivalent in the civil marketplace. However, the learning curve, the efforts needed to understand the requirements of this new marketplace and the expense involved in making contact with potential customers, will almost certainly cost more than the profit on the first few contracts achieved in this area, quite apart from any technical development expenditure involved. Once we have built this bridge the business may be highly profitable, but companies that are not prepared to take the long-term view will never find out.

It is pointless for managers to be bound by internal rules they themselves have set, if these conflict with their business objectives. There are enough problems caused by the competition and the customers – we do not need further problems of our own making.

Another very understandable attitude is that staff joined the company in order to employ their particular skills, as scientists, engineers, specialists in manufacturing, quality etc. They do not expect to be involved in marketing and do not welcome the opportunity. The simple fact is that either they must do it or someone else must be brought in who will do it for them. Services do not market themselves, particularly when times are difficult.

(D) WANTING TO BE PROACTIVE BUT UNWILLING TO MAKE THE COMMITMENT

People in this category know that proactive marketing efforts need to be made, but they try to do it without incurring the cost. One company decided that staff at a certain level should spend 20 per cent of their time marketing. The great merit of this decision was that the salaries were already accounted for. Somehow this one day a week was miraculously going to appear. On reviewing the situation six months later, however, it was clear that the actual time spent on marketing had fallen far short of this figure. Nobody had made the difficult decision that certain activities had to be dropped or delegated or delayed, or that some means should be found of creating the time required for marketing. They had started very enthusiastically but, without much commitment, they found that events rapidly overtook them. The art of management is the ability to control such occurrences.

Another problem in this category is that 'our salary scales will not allow us to recruit marketing professionals'. They either transfer staff without the necessary calibre and experience from other departments, or they set their sights far too low when they recruit. The result is that their marketing is ineffective, and can even become a drain on profit instead of the means of creating profit. Again we have to ask 'who sets the salary scales and makes the rules? What are we trying to achieve?'

Marketing initiatives sometimes come from middle management but are not supported by senior management. Such initiatives are soon stifled. The essence of becoming a market-led organization is that change must start at the top. It is not reasonable for senior managers to expect middle managers to take steps which may change the whole strategic direction of the company without first agreeing with them the objectives, the strategy, the tactics and the resources which will be made available (Follow-up 3.2).

(E) PROACTIVE, PANICKING

Some senior managers, because they have read a book or been on a marketing course, suddenly overreact. They allocate budgets, and recruit or appoint more marketing staff than their business can support at that time.

Investment in marketing should be a planned, sustained and very carefully managed activity, appropriate to the marketplace and commensurate with the legitimate growth aspirations of the operation.

(F) MARKET LED

Some industries, such as those in the fast-moving consumer goods area, have been market led for a long time. The most senior managers are committed to the marketing philosophy, and they generate a climate within their company where marketing professionalism is encouraged and indeed demanded. They spend significant amounts of money on marketing, because they know that this is the only way to generate profit and growth. They set their sights high when recruiting, appointing or promoting marketing staff, realizing that investing in mediocre staff is the worst of all worlds.

Successful companies realize that marketing does not take place only in the marketing department. All senior and middle managers are trained in and committed to the marketing philosophy and are able to play their part in the marketing process.

Another characteristic of a market-led organization is that their business plans are coherent and consistent. They know that they must make an investment before they can expect a return. They have done the arithmetic, described in Section 3.5, which shows how they will achieve a certain level of growth. If necessary, cultures and attitudes are changed and restrictive systems are abolished. Strategic business decisions are based on marketplace considerations rather than on the internal needs of the company.

As a result of professional marketing, such companies are often able to command premium prices and pay above-average salaries. They delegate profit responsibility to capable middle managers, demand results and reward accordingly.

Such an environment is extremely stimulating and fulfilling. Of course, there are problems and frustrations in any business, but at least the directors and managers in this group are taking actions which are most likely to lead to success, and are avoiding the self-inflicted pitfalls of many of the other categories.

CULTURE CHANGE AS A CAUSE OF STRESS

Changing the culture of a company can have far-reaching repercussions at all levels and in all disciplines. It can release creative energy, but it can also cause stress within the management team. Senior managers who lack marketing vision can stifle the efforts of middle managers and lead to immense frustration.

Readers are urged to consider the progression described above, see where they and their company stand in this respect, and decide what they need to do to become more market led. In doing so, they must anticipate the disruption that is likely to be caused throughout the organization. They need to carry their colleagues and their staff with them, so that everyone understands what is happening and why, and what is required of each of them to make it work. The rapid growth in seminars on marketing is an indication of the need for non-marketing staff to be involved in the marketing process (Exercise 3.2).

3.3 THE BENEFITS OF A STRONG MARKETING FUNCTION

Most senior managers are convinced that they need highly professional staff in functions such as manufacturing, research and development, finance, quality, personnel and so on. For some strange reason they do not seem to insist on the same calibre of staff in the marketing department. They transfer people from other parts of the business without any training, or they ask a relatively junior person to deal with brochures, and think that they have set up a marketing department. This dim view of marketing professionalism affects not only the marketing function but also can have equally serious consequences in other parts of the company. Let us assume for simplicity that everyone is either strong or weak, and that we are looking particularly at the interaction between the technical and marketing functions. An example would be a New Product Development Manager who needs a clear marketing specification – a 'user requirement' – before development of a product or service can commence. This can be best illustrated by a simple diagram (Figure 3.1).

Four situations are possible.

WEAK MARKETING INTERACTING WITH WEAK TECHNICAL

This is obviously not worth considering.

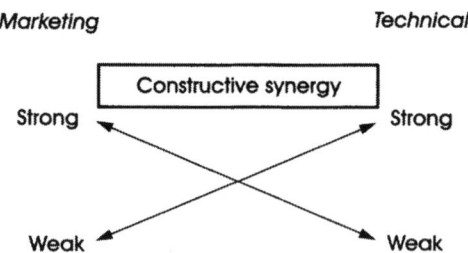

Figure 3.1 The strong–weak interaction

WEAK MARKETING INTERACTING WITH STRONG TECHNICAL

This is a very common situation. It occurs in companies that historically have been technically successful but have not fully appreciated the role of marketing in the future development of their business. They have armies of highly qualified engineers, scientists, research workers, technicians and so on but have not had the vision to match these with marketing staff of comparable calibre. It is much easier in such a company to gain approval for the recruitment of a technical person than the marketing equivalent.

The outcome is very frustrating. Mature technical staff know that their objective is to generate commercially successful ideas rather than to design 'better mousetraps' whose only merit is that they are technically 'interesting'. They know that they should not start a new development project until the Marketing Department has provided an authoritative user requirement specification on which they can base their own technical specification, but they fail to receive it. In its absence, they do their best to make up their own, but they do not have a sufficient understanding of marketplace needs. The only way to find out about customers is to live in the world of customers, not in the R&D Department.

WEAK TECHNICAL INTERACTING WITH STRONG MARKETING

This would be equally ineffective, but rarely occurs. It would mean that the company knew what was wanted but could not deliver it.

STRONG MARKETING INTERACTING WITH STRONG TECHNICAL

This should be the aim of every company aspiring to be market led. A strong marketing function interacting with other equally strong functions can transform a company and be a powerhouse for business development.

Experience shows that this provides a very effective collaborative relationship, and is one of the prime characteristics of an organization which is good at market-based innovation. Good marketing and good technical staff share much in common – they are intelligent, articulate and not hidebound by bureaucratic attitudes – and they can work together very productively. This does not mean that it is 'all one happy family' – marketing and technical professionals may well have deep and protracted arguments – but the result is highly beneficial. The conflict is synergistic and creative, not destructive. Of course, there should be continuing dialogue between the two functions, and I would expect technical staff to challenge very intelligently and vigorously proposals made by marketing staff, but that is not the same as doing their job for them.

In a company selling services, the number of staff involved in marketing or business development may be smaller than in a company selling hardware products, but the need is just as great and the business issues are the same.

The problem is that many senior managers have never worked in a market-led environment. Those who have would never be satisfied with anything less (Exercise 3.3).

3.4 MAXIMIZING PROFIT FROM EXISTING RESOURCES

A common cry is 'we need more people'. While this may be so, we should first see whether we are using our existing people (and other resources) to best effect.

Let us suppose we are offering laboratory testing services to companies running chemical engineering plants. We could do routine testing, send the results together with an invoice, and move on to the next task. This is represented as the lowest level in Figure 3.2. The problem is that a number of competitors, using the same equipment but perhaps with lower overheads, can perform the same basic tests as well as we can. Does this represent a viable business for us if we wish to be a leading contender in our high technology field?

We could redefine our service from 'test' to 'test and diagnose'. Instead of just sending the results, we could also report, on the basis of our experience, that the results indicate that something is going wrong in the process which yielded the samples. It might be a form of contamination, product degradation or a physical defect in the process equipment. In all these cases, what we have revealed might have financial implications much greater than the price of the tests. This could justify our charging a significantly higher price for the more valuable service we are now offering.

But we do not need to stop there. We could again redefine our service as 'test, diagnose and implement'. In the light of what we have discovered, and using our intimate knowledge of the plant and process, we could offer consultancy to introduce changes that can profitably be made. Again, it is

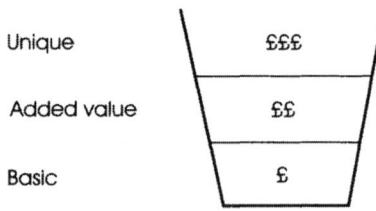

Figure 3.2 Maximizing profit from existing resources

reasonable to expect the client to pay a higher price for such a unique and integrated service.

A challenging discipline is to ask 'do we and our colleagues ever spend our time doing things for which we are over-qualified'? The answer is undoubtedly 'yes'. But how much of our time? In the example above, are we using PhD chemists to perform routine analyses that a less qualified and less well paid person could do? If we allow this question to challenge us to the full, we might even decide that we are pitching the whole level of our business too low, and that we are missing the profitable opportunities which are open to us. We might even decide that we need *fewer* people.

3.5 COMPATIBILITY OF TARGETS AND RESOURCES

One of the signs of marketing maturity is that the resources allocated to achieving the company's turnover and profit targets are compatible with those targets. The number and calibre of staff, and the budgets that they manage, are at the right level for achieving the desired results.

This sounds obvious, but in my experience as a Consultant I often find situations where this is not the case. Managers put up proposals for a certain level of growth, but either do not provide or are not allowed to provide the corresponding marketing resources. This creates a 'time bomb' and one has to ask 'who is trying to trick whom?' The following simple example, and the corresponding exercise at the end of the chapter (Exercise 3.4), will enable readers to check their own situation.

We define:

T = Target annual sales
A = Average contract value
X = % conversion ratio contracts / bids
Y = % conversion ratio bids / prospects
V = value of possible business to be found
N = number of prospects to be found
H1 = person-hours to handle a prospect for which we do not bid
H2 = person-hours to handle a prospect for which we do bid

Let us suppose that we have:

Target annual sales   T = £10 000 000
Average contract value   A = £500 000

We now ask what proportion of contracts we expect to achieve from our bids. Let us assume that this is one in four:

% conversion ratio contracts / bids   X = 25%

Do we bid for every possible prospect we locate? Of course not. There are some we do not want or are not capable of fulfilling, and there are some where we feel that the competition is too intense. Assume that we bid for half of all the possible prospects:

% conversion ratio bids / prospects   Y = 50%

This enables us to calculate the value of the possible business that we have to find

$$V = T \times \frac{100}{X} \times \frac{100}{Y}$$

which in our example is

$$£10\,000\,000 \times \frac{100}{25} \times \frac{100}{50} = £80\,000\,000$$

In other words, we have to locate business to the value of £80 million, of which we will submit bids for £40 million and expect to be awarded the £10 million in our business plan.

Using the average contract size, the number of prospects we have to find is

$$N = V / A$$

which in our example is

$$\frac{80\,000\,000}{500\,000} = 160$$

Let us suppose that it takes an average of 12 person-hours to handle a prospect for which we do not bid (which in some cases may include a client visit) and 120 when we do bid. The total time required is

$$H1 \times N \times (100 - Y) / 100$$

for the no-bid cases, which in our example is

$$12 \times 160 \times \frac{(100 - 50)}{100} = 960 \text{ person-hours}$$

and

$$H2 \times N \times Y / 100$$

for the bid cases, which in our example is

$$120 \times 160 \times \frac{50}{100} = 9\,600 \text{ person-hours}$$

giving a total of 10 560 person-hours.

Assuming 1 600 hours per year per full-time equivalent person, we require 10 560 / 1 600 = 6.6 full-time equivalent people to be devoted exclusively to this task. We have used the expression 'full-time equivalent' people because they may have other responsibilities. If they can only devote 60 per cent of their time to gaining business, we will require 6.6 / 0.6 = 11 people.

This is desperately simple arithmetic, but I meet many people who do not seem to have thought through the implications of their plans even in this basic way. Senior managers demand the results but will not allow the resources. I have also encountered situations where managers are instructed 'from above' to increase their sales forecasts but are not allowed to increase their resources. In fact, if the targets and resources were compatible in the first place, a certain percentage increase in sales may require a disproportionately higher increase in resources – the incremental sales will presumably be harder to obtain. Obviously it is a good management discipline to impose stretching targets, but if they defy logic and are unachievable the result is demotivation not motivation. They are deceiving themselves.

Readers are urged in the strongest terms to judge whether their targets and resources are compatible (Exercise 3.4).

EXERCISES

3.1 Which philosophy in Section 3.1 most accurately describes your organization? If you were the chief executive, what steps would you take to move to a market-led operation? Do you have the right balance between a short and long-term focus?

3.2 Which description in Section 3.2 most accurately describes your organization? What culture changes need to be introduced? What would be the effect of these changes in terms of people, organization and investment? What training is required?

3.3   Consider Figure 3.1. Which scenario most accurately describes your organization? Have you ever experienced the 'strong/strong' interaction? What steps need to be taken to move towards it?

3.4   Insert your own figures into the 'compatibility' calculation in Section 3.5. Are your targets and resources compatible? If not, what needs to be done about it?

FOLLOW-UP

3.1   For the next three months, make a note of the main decisions that occupy your time. Against each, note the degree to which each of these decisions was considered on a market-led basis (e.g. by awarding marks out of 10).

3.2   For the next three months, keep a record of the amount of time you spend dealing with customer issues compared with company issues. Do you think you are spending enough time on customer issues? How can you release more time for the customer (e.g. by abolishing some tasks, meetings or reports, delegating some tasks to others, appointing more junior staff who can save time for senior people etc)?

4

TO WHOM ARE WE SELLING?
MARKET DEFINITION, SEGMENTATION
AND TARGETING

Key business issues   *Section*

O   The definition of the business in which we are engaged
    is not a theoretical exercise but an essential basis for
    business strategy   4.1

O   Historic success does not guarantee that we have a future
    at all, let alone a profitable one

O   The business definition may have to change with time

O   Marketing is making the future happen

O   Customers are not all the same, but they are not all
    different either. Market segmentation defines the common
    factors which link customer groups within each particular
    segment   4.2

O   Segmentation is often defined by industry group, but this
    may not be the most useful basis

○ A number of different segmentation criteria may need to
  be used   4.3

○ Multi-factor or sequential segmentation may be necessary

○ Some key strategic decisions arise from market
  segmentation   4.4

○ Some key marketing actions arise from market
  segmentation   4.5

○ A key to business success is to target our marketing
  activities   4.6

○ Niche markets can have considerable strategic significance   4.7

○ Companies which are most successful at entering new
  markets are those which take the trouble to create
  credibility and track record   4.8

4.1   WHAT BUSINESS ARE WE IN?

DEFINITION OF OUR BUSINESS

Much time and paper is expended on internal documents defining business
mission and related issues. However, these are often theoretical exercises
performed by specialists in staff functions, and they do not have the impact
on line management decision-making that they should.

The only purpose in engaging in such exercises, which take us away from
the task of earning money for the company, is to assist the process of
planning the strategic and tactical future of the business. An example of this
type of exercise is the question 'what business are we in?' In marketing terms,
the issue is fundamental.

THE 'RIGHT' AND THE 'WRONG' DEFINITION

If we had said a few years ago that we were in the typewriter business, the
prognosis would have been disastrous. If, on the other hand, we had said that
we were in the word-processing business, we would have participated in one
of the fastest growth opportunities of the decade.

Similarly, mechanical watches for general use have virtually disappeared
whereas the sale of quartz and other versions has soared. Successful milkmen

sell far more than milk. Petrol stations sell many goods other than petrol, and building societies deal with much more than loans on buildings.

We only have to look at the companies involved in marketplaces under-going similar changes to see how they have answered the question. Some have grasped the opportunity with enthusiasm. Others, for whatever reason, appear to have rejected it (although one wonders whether some of them have even considered the issue).

This need for change is particularly relevant to high technology and fashion industries (interpreted broadly to include, for example, software), where the rate of progress is so rapid and life cycles are becoming so much shorter that companies are constantly having to adjust their focus. The prizes go to those who recognize this fact and react accordingly. Failure comes to those who act as though it will not happen.

Past success does not guarantee a future at all, let alone a profitable one. We might find ourselves appealing to a declining minority of customers. The world does not owe us a living.

THE SCOPE OF OUR BUSINESS

Apart from the actual definition of the business in which we are engaged, a clear statement about the *scope* of our business activities is also important. This is particularly so when we are faced with a changing business environment.

If we define our business too narrowly, we are restricting ourselves and excluding profitable growth opportunities as they arise. Our historic core business may decline, leaving us with an operation which is a shadow of its former self or which even collapses altogether. If we say we are in the defence business, for example, we close the door to the rapid growth opportunities for the same technologies in other industries. If we say we are only supplying the motor industry, we exclude other forms of transport whose needs may be very similar. The relevance of this issue is discussed in Section 4.8.

Faced with changing circumstances, some companies go to the other extreme; they define their business mission so broadly that it does not help them to decide whether or not to engage in a certain activity. I regularly ask my seminar delegates what business they think they are in. I sometimes receive an answer such as 'solving technical problems'. Do they want to repair my car or decommission a nuclear power station, and do they have the appropriate resources!

REACTING TO CHANGE

In extreme examples of a changing business environment, the whole basis of our historic business may virtually cease to exist and drastic restructuring has to take place.

In more normal circumstances, when future trends have been identified well in advance, the art of continuing business definition is to select areas adjacent to the existing business in such a way that they represent an evolution rather than a revolution. We can then proceed step by step, increasing our understanding of the related marketplaces, gradually building a track record, and using this new experience as a foundation for yet further growth. In this way the rate of growth is controlled, the attention of management is not diverted excessively away from the existing business, and any necessary changes in organization or in the quality and quantity of key skills can be managed. If the historic business ultimately disappears, the company is still viable and is in a position to exploit the new opportunities.

By taking this approach, management is able to remain in control. It is no use sitting back and bewailing a drop in orders as many companies do, when confronted with changes in technology, market demand, political or economic factors and so on. The task is to make wise and timely decisions about the issues which *are* within our control. As we have said, 'Marketing is making the future happen' (Exercise 4.1).

Let us illustrate these principles by reference to the nuclear industry. There is clearly a lot of future potential, including decommissioning and environmental work, but there is not enough for all the companies who would like to earn a good living, and there is very little potential for growth except in certain niches. Someone has said that it is like a pond, where the level of water is dropping, but the number of fish in the pond remains the same; to make matters worse, someone has thrown in a few piranha fish!

The first reaction is to say that 'technology is transferable'. We have world-class expertise in research, design, development, project management, standards conformance and so on. All we have to do is to move to some other industries that need these disciplines. The laws of nature are the same in the old and the new markets. An electron does not know whether it is a nuclear electron or a telecommunications electron! The main financial and business issues are generally similar. So what is the problem?

The problem is that the new clients do not only want to be convinced of the scientific capabilities of a supplier; they want to be confident that the client has an intimate understanding of *their business* and can show proven success in *their field*. This vital issue of track record is discussed in more detail in Section 4.8.

This immediately raises a problem. We might, and we probably should, be prepared to invest time and money in gaining an understanding of some new markets to take the place of the contracts that no longer exist in our historic core business. But *which* new markets? How many new fields can we afford to approach? This is where segmenting and targeting is so critically important to our business development efforts.

4.2 MARKET SEGMENTATION

The early chapters of this book have emphasized the need to target our marketing efforts. The reader has been urged to think of marketing as using a sniper's rifle rather than a blunderbuss.

The problem with this approach is that, taken to extremes, it means that every customer is different. If this is so, what is the place of media such as brochures, advertisements, web sites, videos and so on, which need a reasonably large audience for their financial justification?

The answer to this dilemma lies in market segmentation. Customers are not all the same, but they are not all different either. The need is to find a common factor, or factors, which links a particular group or segment. We can then design a marketing and sales approach which is uniform for that particular segment but which is different from that for the other segments.

Let us suppose that we are selling computer software to industry. The obvious way of segmenting the market might be by industry classification, e.g. financial, retail, manufacturing and so on (see Figure 4.1).

However, a little thought may show that the needs within a particular industry segment such as finance, measured by some factor such as size, sophistication or purchasing power, vary widely. At the top end (F in Figure 4.1) – the headquarters of an international bank – the sale may require months of negotiation, detailed proposals, presentations, meetings and so on, and the result will be an individually specified package. At the bottom end (f) – a small accountancy practice – the sale may be achieved in one hour by a visit from a sales representative, or even by sending a catalogue or brochure, and the customer buys an 'off the shelf shrink-wrapped' package. If we then look at another industry segment such as retailing, exactly the same can be said. We begin to wonder whether the segmentation should, in fact, be at right-angles to the segmentation by industry, because there is much more in common *between* segments at a particular level than there is *within* each segment. In our example, there is much more in common between the computer software requirements of the international bank (F) and a large

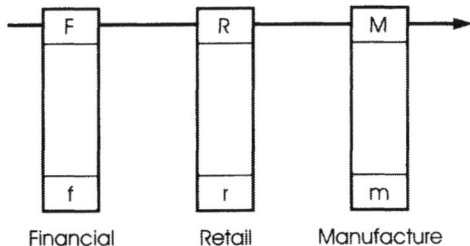

Figure 4.1 Industry segmentation

supermarket chain (R), which are in different industry segments, than there is between the bank (F) and the accountant's office (f) even though these are both in the financial sector.

On the other hand, we do need to consider industry segments as well, so that we can show that we understand their particular business and the ways in which it is different from every other business. In our example, we need to demonstrate that we are familiar with the language, culture and jargon of banking and finance, and that we can relate to their business environment, quite apart from any technical aspects of their specialization (Exercise 4.2).

We are therefore driven to the conclusion that there is no single segmentation criterion that adequately serves all purposes. The objective is to arrive ultimately at one or more segments that can be approached with relevant and targeted messages with a dedicated marketing programme.

4.3 SEGMENTATION CRITERIA

What are the possible criteria or factors that we might use to segment a marketplace? Three ways of approaching it are suggested (but there are many others which readers can devise for themselves):

1.   *The type of customer.* The criterion might be the location of the customer. This is particularly true if we are a relatively small operation for which it is not economic to go beyond certain geographical boundaries. It might be the size of the organization – below a certain size they would not be interested in what we had to sell or the level of business we might achieve would not justify the cost of selling to them. Conversely, above a certain size, the competition might be so intense that we would be better to concentrate our efforts lower down the scale.

     The segmentation criterion might be the profitability of the organization or the wealth of the individual, particularly if we are selling a sophisticated high-priced service. A further distinction might be made between existing and new customers, for which an entirely different marketing approach will be required.

2.   *Type of assistance needed.* Segmentation might be by tangible factors such as quality, performance and continuing support required, but there may also be an influence from intangible factors such as brand preference, 'we always (or we never) use one of the top world-class consultancies'.

3.   *The way purchasing decisions are made.* The segmentation criteria might be related to the amount of outside consultancy traditionally

used, the influence of price on the decision, whether the decision to use outside consultants is centralized or delegated, and whether it is made by one person or a group. These criteria will determine the way in which we sell the same service to different segments.

To summarize, market segmentation is not a theoretical exercise. There are only two reasons for doing it – so that we can make decisions (Section 4.4) and take actions (Section 4.5).

Because there is no single universally applicable segmentation criterion, we may have to make a multiple or sequential segmentation. Let us assume that we are a high technology resource in a large international process industry; our prime customer is our own company, but we are also being encouraged to seek business in outside companies. We now apply the three criteria described above. Figure 4.2 shows the result of defining four, three and three segments respectively within each step.

This gives 36 (4 × 3 × 3) possible combinations such as:

O   Own company large / State of the art / Committee
O   Own company small / Regular support / Department managers
O   Outside company small / *Ad hoc* projects / High-level centralized.

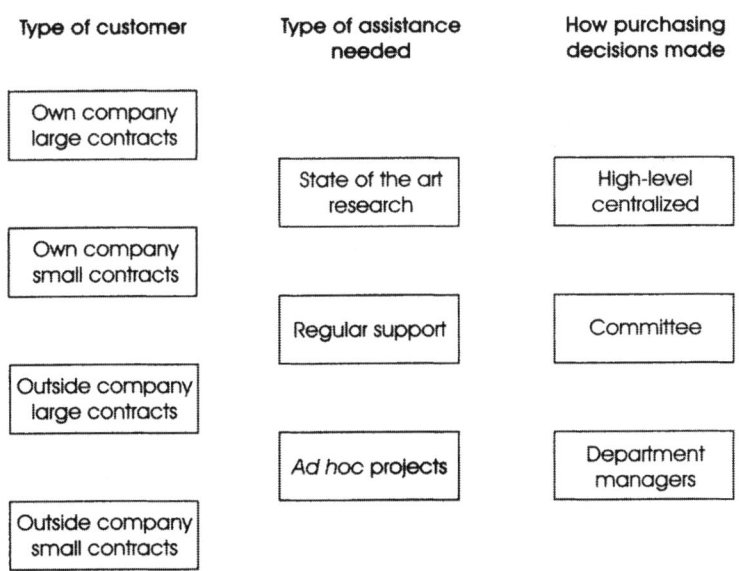

Figure 4.2   Sequential segmentation

Some combinations, such as Outside company small / State of the art / Department managers, are unlikely to exist and can be eliminated, but that still leaves a bewildering number of strategic options.

There is nothing fixed about the number of steps or the number of segments within each step. The technique does not, of itself, tell us which if any of the segments we should approach. In this example, if we put all our resources into only one of the 36 segments so defined, we would be eliminating a large part of the target market. If we tried to target them all, we would over-extend our resources and not hold a leadership position in any (even in our own company, because outsiders are aggressively encroaching on what we had regarded as our own market share). In practice, we would probably select some compatible combinations, giving us a reasonable market size but retaining a good degree of targeting.

It is as dangerous to be excessively targeted as it is to dissipate our efforts in all directions. This is a serious business dilemma, but it does not become any easier if we ignore it (see Exercise 4.3).

4.4 STRATEGIC DECISIONS ARISING FROM MARKET SEGMENTATION

Possible decisions arising from market segmentation include the following.

WHICH SEGMENTS SHOULD WE CHOOSE?

Some companies deliberately choose only one market segment and concentrate exclusively on that. Others may decide to target all possible segments. A good approach might be to concentrate on a small number of segments at first, and then move into others as our position becomes established in the first segments. There is no 'right answer' to this question, but whatever we do should be the result of a conscious decision after weighing up the alternatives, rather than allowing it to happen by default because no one has taken the trouble to think through the implications. The issue of targeting is discussed in more detail in Section 4.6.

DO WE NEED A DIFFERENT ORGANIZATION?

One of the possible segmentation criteria described in the previous section was the way in which purchasing decisions are made. These can vary so radically between different market segments that it may well be necessary for us to reflect this in the way we organize our sales and marketing efforts.

At the 'top' end of the market, we will need to employ senior skilled negotiators and, probably, proposal writers. At the 'bottom' end, a team of traditional sales representatives may be the most appropriate. People who are good at one may not necessarily be good at the other, and it may be necessary to recruit in order to buy in the necessary skills and experience.

Individual members of the senior management team may have a key role in selling. The chief executive, the financial director, the operations director and so on may be the best people to sell to their counterparts in the client company. If this is so, their workload must allow them to have time for customers. I was once asked to run a session on marketing for the board of a large subsidiary of one of the world's largest companies. I assumed that they wanted the latest theories on marketing from international gurus. I was then told that what they actually wanted was a session on 'selling techniques for directors', because they clearly recognized the point we have just made.

ARE WE PREPARED TO MAKE THE INVESTMENT?

Existing sellers will almost certainly have created barriers to entry to their segment. We may have to spend time and money before we can expect to make the first sale in what to us is a new segment. Such an investment may well be worthwhile, and establishing our place in the new segment may safeguard the future of our whole operation, but we should count the cost before taking the first step.

This approach contrasts strongly with the philosophy of dabbling in the whole market and responding reactively to opportunities that happen to come our way. This is not to say that we should reject opportunities that arise spontaneously, but relying on this approach does not constitute a professional and proactive business strategy.

WILL DIFFERENT SEGMENTS NEED A DIFFERENT SERVICE OR VARIANT?

Life is obviously easier for us if we can sell an identical service in all segments, but this is rarely possible. To attempt to do so is to limit our chances of success from the start. The customers in each segment need to feel that the service matches their particular requirements, while we want to incur the minimum of work and expense. We need to find the point where the customers perceive maximum specificity while we achieve maximum commonality.

CAN DIFFERENT SEGMENTS BEAR A DIFFERENT PRICE?

The normal approach to pricing in much of industry is to calculate the cost and add a percentage. The fallacy of doing this is argued in Chapter 6. The marketing approach to pricing is to judge what price the market will bear, based on the perceived value of what we are offering.

It is inconceivable that all market segments will value a service equally. The reasons why a service is needed will vary between segments, as will the value it can provide, the degree of competition and the cost of alternative solutions or of doing without the service, and these all have an influence on price.

For these and other reasons, we will be most successful when we most accurately judge what price each market segment will bear for our services. The result will be the mix of high and low margins and volumes which gives us the maximum overall profit.

The role of segmentation in decision-making is yet another example of the need to target every part of our business activity to the particular needs of the marketplace (Exercise 4.4).

4.5 MARKETING ACTIONS ARISING FROM SEGMENTATION

Possible marketing actions arising from segmentation include the following.

SHOULD WE CLAIM DIFFERENT BENEFITS?

Different people want different benefits from the same service, and something that is a benefit to one person may be a dis-benefit to another (see Section 2.4). We need to think through the implications of this for our entire marketing platform.

SHOULD WE PRODUCE DIFFERENT FORMS OF COMMUNICATION?

Taken to its logical extreme, we will have to produce a different version of our literature and other forms of communication for each segment. Literature that tries to be all things to all people ends up not being very much to anybody.

Imagine a high technology service that could be used in a number of industrial situations, such as routine manufacturing, field service, research, avionics and telecommunications. We might produce a series of five brochures. Much of the material, covering the details of the company and the general features of the service, would be common to the five versions. We could leave room for a photograph with which each individual target segment could identify and a panel that highlighted the main benefits for that particular segment. By cleverly designing a family of brochures, we would be able to cover five very different targets for about the cost of two entirely separate brochures. Creative thinking of this sort can achieve both increased impact and reduced cost.

Exactly the same thinking can be applied to other forms of communication, such as advertisements, web pages, overhead presentations, 'standard letters' or whatever is appropriate.

SHOULD WE ADVERTISE IN DIFFERENT MEDIA?

The most effective forms of marketing communication are normally those that use highly targeted media. Generalized advertising claims in non-specific media are only appropriate if we are selling to a very general readership, which is not the case for most high technology services.

The cost per subscriber of the advertisement will almost certainly be higher in the more specialist publications, and this has to be borne in mind when deciding media strategy. However, the only effective measure ultimately is the cost of reaching a new customer or, more correctly, the cost of achieving £1 of net profit as a result of the advertising.

If we decide to advertise in a number of different media to reach different market segments, we need to target the message. As discussed above, we may only have to change a few key words or a picture, but it can make all the difference to the impact.

SHOULD WE OFFER DIFFERENT SALES SUPPORT?

This could include areas such as training, technical support, problem-solving, after-sales service, enhancements, software upgrades and so on. The degree to which these areas of sales support are required will vary with the segment.

Again, the more we can improve our interaction with customers in each segment, the more credible will be our presence in the marketplace and the more successful we will be (Exercise 4.5).

4.6 TARGETING

Segmentation is an important marketing issue because none but the largest companies can profitably address all segments of a market. Targeting is the application of our marketing efforts specifically to the segments or sub-segments we have chosen.

We need to recognize that our targets may be changing, otherwise we may be concentrating on the wrong target. This happens particularly in the fast-moving markets typical of many high technology businesses. Consider the market for computers and all the services that go with them. Originally most of the turnover was in mainframes. Next came workstations and minicomputers. These markets then subdivided into portables, laptops, notebooks and so on. The companies which have been most successful in this fast-moving field are those which have pioneered the changes. Companies that have tried to live in the past have suffered the consequences. On the other hand, being too early can be as disastrous as being too late.

Marketing strategy must include an evolving targeting plan. We should not only be planning to target the existing segments but be thinking ahead to the next one or even two stages in the developing marketplace. A useful analogy is that of a military commander setting out to invade a country, who establishes a beach-head, consolidates it, moves forward into a growing series of strongholds and gradually plans to take over the whole territory. At each stage, everything is done to ensure that the enemy cannot retake lost ground (Exercise 4.6).

SELECT, FOCUS, TARGET AND COMMIT

We need to select and focus on certain market segments or sub-segments, target our marketing efforts on the ones we have chosen, and commit appropriate resources in order to achieve our specific goals for these targets. This approach will give us a level of success which unfocused, untargeted and uncommitted marketing efforts can never achieve (Follow-up 4.1).

4.7 THE STRATEGIC SIGNIFICANCE OF NICHE MARKETS

We have spoken of segments and sub-segments. A still further narrowing down of a sub-segment can be described as a 'niche' market.

At first sight, a niche may appear to be unimportant because of its small size. However, handled correctly, a niche can offer a highly profitable marketing opportunity to a particular company, and can permit the creation of such barriers to entry that competitors may decide to leave it alone. This is the ultimate objective of niche marketing – one company not only dominates the niche but virtually 'owns' it.

As with segments and sub-segments, niches may exist on a continuing basis, or they may come and go in a rapidly changing situation. The skill lies in identifying, capturing and exploiting such niches and then, if necessary, moving on to others. Figure 4.3 illustrates the point that what we think of as one marketplace is actually made up a number of different markets and niches.

The markets and niches are defined by the variables along the two axes of Figure 4.3, based on customer segmentation and product segmentation. As an example, customers for training in marketing and business planning might be segmented into:

O   companies experiencing increased competition
O   companies diversifying into new markets
O   companies requiring their R&D to become more market led
O   companies wishing to set up a new business development department

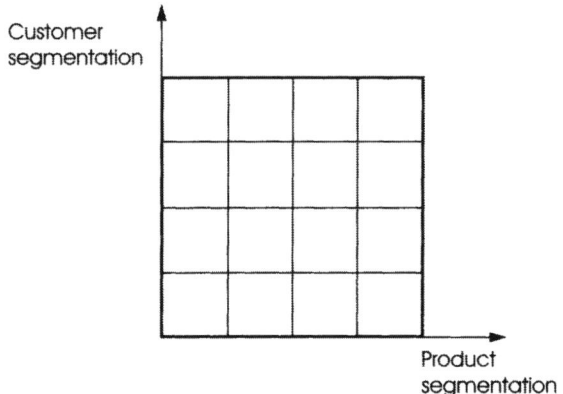

Figure 4.3   Niche markets

O   companies wishing to use fee-earners for marketing
O   companies needing to improve their customer service
O   organizations moving from being reactive to being proactive, and so
        on.

The products or services that such different customers require might be:

O   individual places on public seminars
O   in-house standard seminars
O   in-house tailor-made seminars
O   guidance on writing business plans
O   consultancy assistance with writing business plans
O   consultancy assistance with approving business plans, and so on.

Having defined Figure 4.3's two axes, we obtain a number of 'boxes'. We then write the names of companies supplying to each particular niche in the appropriate box. An estimate of the turnover achieved by those companies would give added value to the exercise.

The probable outcome is that we shall find a cluster of companies selling in certain customer/product areas. The obvious inference is that the greatest potential is where the largest numbers of companies are operating. However, a little more thought shows that these are the very areas where competition is greatest. It may be possible to identify one or more smaller market opportunities which competitors have not begun to exploit. Although the potential is less, it may be much easier to achieve a high share of one of these markets than 1 or 2 per cent of the busiest markets; the resulting turnover and profit may be higher, and the marketing costs lower.

For some companies, a single well-chosen niche may be adequate to fulfil all their ambitions; it has the great advantage that the larger companies will not feel threatened and will probably not take any steps to retaliate.

Other companies may not feel that one niche market offers sufficient potential to satisfy its long-term needs. However, it may represent an excellent 'stepping-stone'. This is particularly true if a company is venturing into a market outside its traditional core business, where it might not be appropriate to confront the existing market leaders head on. A flanking attack, by which it can establish a foothold, test the acceptability of its service, test its marketing tactics and confine any mistakes to a level where their impact is not serious, may be an excellent prelude to a stronger attack on the more substantial parts of the marketplace (Exercise 4.7).

In my own case, I have carried out a three-stage niching process to define my business priorities. First, I could say that I am in the general business of management consultancy, offering a wide variety of expertise. I do not do so; I restrict my activities to marketing and business planning. Second, I could offer my services to a wide range of industries. I do not do so; I only work with high technology companies. Third, I could offer consultancy of any type. I do not do so; I only undertake consultancy where it is integrated with an in-house training programme. In this way, I have established a very specific focus, which has a number of very great business advantages.

Most readers will not work for one-person consultancies as I do, but the principle is exactly the same in a larger organization. It is much more credible to offer multiple specialisms than to try to promote general practice.

4.8 CREDIBILITY AND TRACK RECORD

THE IMPORTANCE OF TRACK RECORD

Successful companies have created a track record in their core business areas. This is an immensely valuable asset that brings them continuing business.

The problem comes when such companies need to move into new areas where they are relatively unknown. A delegate on one of my seminars said wryly, 'we tried to start supplying the railway business, but unfortunately we didn't have a track record!'

Examples are companies whose historic core business has been in areas such as defence, nuclear and space. These markets will continue to provide business for the lifetime of the companies concerned, and it is right that their first priority should be to 'milk' the markets they know best. However, the demand has plateaued and begun to decline, while the number of companies hoping to earn a living from of them has remained more or less constant. In these circumstances, companies wishing to continue to expand, or even to maintain their traditional profitability, will have to find new markets in which to ply their expertise.

They might think that all they have to do is advertise and the business from the new area will start coming in. Unfortunately this is not so, and companies who have tried it have learned the lesson the hard way. The failure is predictable, and it is a failure to appreciate the marketing issues involved. It arises from insufficient understanding of the mind of the customer.

Buyers consider a number of potential suppliers, perhaps starting with a 'long list', moving to a 'short list' and ultimately selecting one. A supplier who is not on the initial 'long list' will never be considered. One of the most important determining factors is the perceived track record in that particular area of business.

Let us suppose that we are looking for project management skills for a large water distribution project. Would we seriously consider suppliers to the defence, nuclear or space industries who know little about the water industry? Why should we take the risk of the unknown, when there are large and experienced suppliers who have already worked in our industry, who have established a good reputation and who can quote a long list of successful and relevant projects. The would-be supplier from another field has a low or even zero profile in the new area.

The position may actually be worse than having a zero profile – the 'outsider' may in effect have a negative profile. The technical criteria in the water industry may be somewhat different from those in the defence, nuclear or space industries. The prospect might hold the view, rightly or wrongly, that a contractor from one of these ultra-high technology industries would be likely to offer over-engineering, a bureaucratic process, excessive time scales and a high price. It is against this background that would-be suppliers have to begin their marketing. (This argument has nothing to do with environmental issues, although these might represent a further obstacle.)

HOW CAN WE CREATE AND DEMONSTRATE A TRACK RECORD?

One of the main obstacles to be overcome in moving into a new business area is to win the first contract. The problem is, of course, that we have no record of success on which to base our claim to credibility. It is much easier to win contract number 5, 10 or 100, and yet the whole process cannot start until we have gained the first.

These early successes are so important that it is worth going to great lengths and, if necessary, incurring substantial expense in order to achieve them. This does not necessarily mean offering a low price, because this can negatively affect our image and make it difficult to raise prices later, although 'buying some business' may sometimes be an appropriate strategy. It might be better to 'overkill' the opportunity by bringing to bear considerably more technical, marketing and management attention than will be necessary once the track record has been established. Even if the early contracts are not particularly profitable in their own right, they form an essential stepping-stone towards the development of a continuing business.

Customers think, with some justification, that their business is special and different from all other businesses. There are matters such as language, jargon, culture, ways of doing business, and other factors, which we ignore at our peril. To move from where we have been to where they are may require a deliberate programme on our part.

How do we create the necessary credibility? We may have to recruit someone who has worked for many years in the new market segment, who can 'live, breathe, eat and speak' the culture of that particular industry. We could use such a person first to identify opportunities and problems in the new segment and, second, to spearhead our marketing and selling approach in that market. If this were not appropriate, we might second a member of our existing staff to do nothing but concentrate on that segment for, say, six to twelve months. The brief would be to identify the opportunities and evaluate the implications of entering the new segment, leading to a business plan that would then be submitted for approval or modification. Once approval had been given, he or she would be the logical person to mastermind at least the initial entry into that market while additional resources were being assembled. As a side benefit, a secondment of this type could be a very good way of developing the management skills of the person selected (Exercise 4.8).

AN IMAGINARY CASE STUDY

The 'case study' below is a composite picture based on the combined experience of a number of companies wanting to diversify out of their traditional areas. No reference to a particular company is intended.

The starting point is that the companies concerned had all been commercially successful and had reached positions of leadership in their field. They expected to continue to make reasonable profits from their historic core business, but the demand was falling and was insufficient to meet the aspirations of all the suppliers. To generate growth, they had to enter new markets of which they had little previous experience. Their management and marketing resources, which had been tailored to the earlier situation, had to undergo a major restructuring.

The problems facing our imaginary company in the new situation were as follows:

1. *Management* Their management style was unduly conservative, bureaucratic and averse to taking risks. Staff showed the wrong business culture and attitudes.

2. *Company name*   The company had the wrong name and the wrong image.

3. *Marketing*   The company had little orthodox marketing experience or expertise. Business had historically come to them from other companies with whom they co-operated, or had been gained as a result of government tendering procedures.

The Marketing Department was too small, was populated by low-calibre staff and had little responsibility outside the areas of tender preparation and brochures.

There were no sales staff, and managers had few contacts outside their core business area.

They had no separate marketing budget, and tried to 'hide' marketing expenditure within existing project codes.

They had the wrong literature, and a corporate video that was entirely inappropriate outside their core area.

4. *Pricing*   Pricing was based on cost and overheads, with no understanding of the price which the new markets were willing to bear. Their historic business had a much higher level of overheads than was appropriate to the new business areas they were seeking to enter.

5. *Business planning*   They had no agreed business plan to which everyone was committed, and their business development efforts were fragmented and inconsistent.

Faced with the above situation, the company took the following actions:

1. *Management*   They made changes at senior and middle management level, filling some key posts with managers who had held responsible positions in highly competitive market-led outside industries.

They delegated responsibility to middle management, encouraged investment and risk-taking, and then demanded results.

2. *Company name*   They changed the company name to one which did not solely reflect their former business area.

They took steps to change their image as perceived by the outside world, presenting themselves as 'fast-on-the-feet' operators who could make decisions quickly.

3. *Marketing*   They strengthened the marketing function by bringing in some key managers from professional marketing environments, and undertook a major programme of marketing

and business training for the existing key staff.

They created and allocated marketing budgets.

They set marketing objectives that were reviewed after six or twelve months. They appointed full-time sales staff and introduced sales incentives.

They created new literature for each of the target markets, in a flexible portfolio form that could be updated as the track record developed. They avoided using the generic company literature and corporate video when they felt that these would give the wrong signals.

They set up a proactive marketing programme that included mailshots and 'cold calling' into new areas.

4. *Pricing*   They agreed a more flexible approach to pricing, based on customer value (which involved some tough battles with their corporate senior and financial management). They eliminated some areas of overhead which were not appropriate to the new circumstances.

5. *Business planning*   They segmented the potential new marketplace, thought through the implications of addressing the various segments and developed an 'invasion plan'. This involved prioritizing a small number of targets for initial marketing efforts, planning a second phase expansion into further areas, but deliberately avoiding dissipating their efforts over too wide an area too soon. They appointed business managers for the main target market sectors. These were senior managers of high calibre and experience, who were given responsibility for the profitability of the activities relating to the new areas (although they used common technical and other resources as appropriate, to avoid fragmentation and duplication). In particular, they were responsible for all sales and marketing activities in their own marketplace (see Section 12.2).

They recruited some key staff from the main market sectors they had selected.

They then invested time, money and resources in creating a track record in the new market sectors. They recognized the strategic importance of the first contracts in these new areas, and were prepared to go to great lengths to obtain them; profitability of these first contracts was not the top priority.

They wrote a business plan (using the framework shown in Section 12.4), which was approved by senior management and which became the guide to their business development efforts.

The management teams of the companies represented in this composite picture deserve credit for the seriousness with which they are building a new business and for their willingness to take a long-term market-led approach. They are already achieving considerable success, and their company culture has been transformed beyond recognition. Far from demotivating scientists and engineers, the new commercial environment has given them a new sense of optimism and purpose. This strategy is to be contrasted with that of other companies who sit back and complain and feel that the world owes them a living (Follow-up 4.2 and 4.3).

EXERCISES

4.1   Define in not more than 20 words the business in which you think your organization is (or ought to be) engaged. Are you satisfied that the definition sufficiently reflects future trends in the marketplace? Are you satisfied that the scope of the definition is not so narrow that it precludes growth or so broad that it does not help in diversification decisions?

4.2   Write down the way in which you normally segment your company's marketplace at present. In the light of Section 4.2, are you sure that this is the best way of doing it?

4.3   Carry out a sequential segmentation exercise for one of your main areas of activity. What issues does this raise which are not raised by a one-stage segmentation?

4.4   What strategic decisions do you need to make as a result of your increased understanding of market segmentation?

4.5   What marketing actions do you need to take as a result of your increased understanding of market segmentation?

4.6   Write an evolving targeting plan for your operation to cover at least the time span of the long-range strategic plan.

4.7   Fill in the grid shown in Figure 4.3 for your own operation. What niche marketing opportunities does this suggest?

4.8   Choose a market that is new to your company which you are considering entering or taking more seriously than in the past. What steps should you take to create and demonstrate a credible track record? Should you appoint or second someone to 'champion' the opportunity?

FOLLOW-UP

4.1   Over the next six months, make a list of any marketing activities in which your company engages which are unfocused, untargeted or uncommitted. At the end of the period, share your findings with the relevant staff and decide what action needs to be taken to overcome these deficiencies.

4.2   Select a small group of colleagues, and make them familiar with the imaginary 'case study' in Section 4.8. Arrange to meet regularly, say every two weeks for a period of three months, perhaps over a sandwich lunch, to explore its relevance to your operation. At the end of the exercise, write a brief report to senior management with some clear and supported recommendations.

4.3   Study the business planning framework in Section 12.4, and consider whether to adopt this approach.

5

WHO ARE OUR COMPETITORS?
COMPETITIVE ANALYSIS AND TACTICS

Key business issues	Section
○ Competition may come from a variety of sources, many of them not the obvious ones	5.1
○ Many companies wrongly act as if the only competition were from other suppliers of an equivalent service	
○ Competition may come from other ways of doing the same thing, an alternative service, other demands on budgets or from customers doing it themselves	
○ Failure to analyse the competitive situation may mean that the wrong battle is being fought	
○ A detailed and continuing analysis of the competition is a key part of the marketing function, and may reveal unexploited marketing opportunities	5.2
○ There are many sources of information on the competition, some of which are often under-used	5.3

○ Competitive tactics are a key part of the marketing
   function   5.4

○ Attacking the industry leader carries particular risks but
   it can be successful if certain principles are followed

5.1 THE NATURE OF COMPETITION

A study of the competition is a fundamental part of the marketing task, and
most of this chapter is devoted to ways in which we can improve our performance against other companies providing broadly equivalent services.

However, many people do not think sufficiently laterally about the nature
of competition. As a result, they employ the wrong tactics in fighting the
competitive battle. Most of their marketing effort is directed at showing why
their service is better than the others on a like-for-like basis. As we shall
show, competition can be very much wider than this and can come from
entirely different sources. Failure to recognize this can mean that we win the
argument but fail to win the contract because we are arguing the wrong
issues.

Let us illustrate the point by supposing that we are offering chemical
laboratory testing services for process control. There are five possible forms
of competition:

1.   *Other providers of the same service.* These will be other suppliers of
        chemical laboratory testing services. Many of them will be using the
        same equipment as ours with results which we have to admit are as
        good as ours. We have to find some way of showing why our service
        with its associated benefits is to be preferred to the other services –
        by having a very efficient sample collecting service, for example.

2.   *Other ways of doing the same thing.* Alternative methods could
        include a laboratory that measures physical rather than chemical
        properties. Already the competitive ground has shifted. We are no
        longer trying to show why our chemical analytical service is better
        than the other chemical analytical services. We have to sell the advan-
        tages of chemical over physical testing.

3.   *Providers of an alternative service.* An alternative service might be a
        system for in-plant process control. We could be successful in
        proving that ours is the best external laboratory service, and the
        plant operator might say that, if they wanted external laboratory
        testing, they would use our services. Unfortunately, they prefer to

have in-plant process control. In this case, the competitive battle we have to fight is, first, to show why external testing is to be preferred to in-plant control and, only then, to argue why our laboratory is the one they should use.

4.   *Other competitors for the customer's budget.* Some plant managers may think that the cost of testing is not justified, and they are prepared to accept the process as it is. In this case, we need to show why money is better spent on external laboratory services than on something entirely different, such as external maintenance or catering services. Put more generally, if a customer would like to have a number of things and can only afford a few, we must aim to make sure that our service is one of the few. The competitive factors bear no relation to the high technology we are offering.

5.   *The customer.* It is a startling fact that the customer can be our biggest competitor. In our example, the plant manager might set up a simple in-house chemical laboratory. In this case, our task is to argue the business benefits of our service, which gives more reliable results, better process control and perhaps a better product at the end of it. A word of warning – we might think that the customer will do a bad job, and we may well be right, but we are not very likely to win the contract if we say so!

These examples can easily be transferred to our own selling. The point is to identify the actual competitive scenario in each case, so that we can bring to bear the relevant competitive tactics. If we ignore this, we can 'win the battle and lose the war'.

We need to think laterally about the competition (Exercise 5.1).

5.2 ANALYSIS OF COMPETITION

Having made the point that competition comes from a variety of sources, we now revert to considering the simplest form of competition – other suppliers offering broadly equivalent services.

A knowledge of the competition is an absolutely essential part of the marketing function. The task is made harder by the fact that circumstances can change rapidly and frequently. It is a hallmark of good marketing managers that they have an accurate and up-to-date understanding of the strategy and tactics of the main competitors. What sort of information do we require?

1.   *Who are they?* This should include an understanding of their owner-ship and organization.

2.   *What is their financial situation?* Are they fighting for survival? Do they have surplus funds to invest in new ventures? Are they able to increase their expenditure on marketing? Are their salary scales competitive – are they in a position to attract key staff from their competitors (including us)?

3.   *What services are they selling?* Competitive brochures and other key information should be available, kept up to date and in an accessible form.

4.   *At what prices are they selling?* Competitive price lists should be available if they exist. Even more importantly, there should be an understanding of the degree to which their fees are negotiable, and their approach to pricing tenders, if this information can be acquired.

5.   *What is their market share by segment?* A broad idea of competitive market shares is essential to developing our competitive strategy. As will be discussed in Chapter 10, the cost of obtaining such informa-tion increases very rapidly with the accuracy required. The key question is 'if we knew the figures more accurately, would it influ-ence any business decisions we might make?'

6.   *What product features are they promoting?* An analysis should be made and kept up to date of all the main features, specifications and claims of each of the significant competitive services compared with our own.

7.   *What benefits are they claiming?* Not only should we sell benefits rather than features – we should present our benefits as being more beneficial than those claimed by the competition.

8.   *What are the strengths and weaknesses of the competition?* It is a useful exercise to assess the relative strengths and weaknesses of our own company and the main competitors. This is normally done as a SWOT analysis – see Section 12.8. We are, as it were, playing a game of chess; we need to devote at least as much effort to the serious matter of business competition as chess enthusiasts would devote to what is, at least in theory, only a game!

9.   *What is their retaliation potential?* Consider what actions main competitors would take if we were to do certain things. If we are the

market leader, we may decide that we are relatively invulnerable; if we are not, we should be aware that a competitor might take steps which could seriously damage our business or even drive us completely out of the market.

The reader may be able to add other issues to this list.

To summarize, the marketing process does not relate simply to the buyer and the seller; the competition can be a significant factor in the equation (Exercise 5.2).

5.3 SOURCES OF INFORMATION ABOUT COMPETITION

How can we obtain detailed and up-to-date knowledge of the competition? We are not advocating industrial espionage, but we do need to make a continuing appraisal of the competition. This is rather like creating a jigsaw; each piece of information may not be highly significant in its own right but, over a period of time, a number of pieces combine to make a very useful picture.

1. *Published sources.* Significant pieces of information are published from time to time, in business, financial, technical and other journals, in the newspapers, on television and so on. We need to set up a system for capturing and accessing this information, and for maintaining it on a regular basis. It is worth devoting considerable amount of expert effort to the system design, but the maintenance can be carried out by carefully briefed relatively junior staff. Data can be computerized with key word access, or it can be stored in paper filing systems with a good method of reference.

2. *Customers.* Customers are in frequent contact with our competitors. They receive early indications of new services, price changes, changes in methods of operation, special promotional and other activities, and so on.

3. *Sales force or fee-earners engaged in selling.* Even if fee-earners do the selling and we have no separate sales force, such people are in daily contact with the marketplace and are constantly receiving snippets of information about the competition. The issue is whether this information is reliable and statistically significant, and how it can be incorporated into a broader picture.

    The simplest starting point is for management to insist that every daily or weekly sales report and every overseas visit report should have a section on Competitive Activities. Also, sales staff should be encouraged to telephone 'hot news' to Head Office. The wise manager will not over-react to the first report of a certain competitive activity – it might simply be a salesperson's excuse for having lost a contract! However, when we begin to hear the same thing from several different sources, we can assume some measure of statistical significance.

4.   *Using competitive services.* The use of competitive services is not, of course, always possible, but it can sometimes be done as a means of assessing the competition. Similarly, we may be able to learn something if we use competitors as subcontractors.

5.   *Exhibitions.* Exhibitions are a significant and instant source of information. When planning the staffing levels for an exhibition stand, it is worth allowing time for a detailed study of the competitors' stands. At the end of the exhibition the staff should be brought together to pool the knowledge they have gained. The latest competitive literature should be gathered and studied after an exhibition as an aid to developing our own literature (see Section 9.1).

6.   *Independent market research.* Both qualitative and quantitative studies can be commissioned, on a one-off or regular basis (see Chapter 10). As with all market research, the criterion is whether the information we gain is cost-effective in terms of enabling us to make better business decisions.

7.   *Interviews and recruitment.* In many industries, interchanges of personnel between competing companies are frequent and they will bring facts and attitudes with them.

The better we are at harnessing the information from these various sources, by setting up a really workable competitive database, the better equipped we will be to fight the competitive battle (Follow-up 5.1).

5.4 COMPETITIVE TACTICS

FORMS OF COMPETITION

The competitive situation can vary from a monopoly at one end of the scale to a totally free market at the other. Most high technology service companies are operating somewhere between the two extremes, probably nearer to the free market.

In a monopoly there are no competitors; the monopolist controls the supply and, to a large extent, the price. Some monopolies exist because of legislation or because the industries are government owned – some public utilities, for example. Where such a monopoly is considered detrimental to the customer, a government sometimes steps in to create a free market.

Other monopolies or near-monopolies come about because a large supplier establishes such a dominant hold on the marketplace that it is very difficult for others to enter.

As we move from a monopoly towards a free market, there is an increase in the number of competitors and the number of different services being sold, and a lowering of the barriers to entry. At its extreme, a free market would be described as offering 'perfect competition'; there would be a large number of sellers, none of whom could control the supply or the price.

The main issue, in whatever competitive situation we find ourselves, is 'what reasons can we create for customers to buy from us rather than from the competition?' How can we differentiate ourselves in such a way that customers will put us at the top of their shopping list? In a sense, the free marketeer is continually trying to create some of the advantages that the monopoly supplier naturally possesses, by attempting to 'own' the chosen segment or niche.

Dominant companies will usually lead the marketplace and be first in the field with new services. Advantageous though this strategy may seem, there is an argument, particularly for smaller companies, to follow rather than to lead. This has the merit that the leaders can be allowed to spend the money to create the market and, sometimes, to make the mistakes; followers can then come in at a much lower level of effort, risk and expenditure. This has been described as being 'first in the field with a "me-too" service'; being tenth in the field with a 'me-too' service is not such a good strategy!

ATTACKING THE INDUSTRY LEADER

What do we do if there is a clear industry leader who dominates a market segment in which we are already operating or which we are considering entering?

It must be said that industry leadership is a very powerful position. The knowledge, experience, service range and financial investment of the market leader can form a serious barrier to competitors. Does this mean that we should never contemplate attacking the industry leader? Certainly not, but the process carries a high risk and we must carefully think out our tactics before embarking on such a venture.

Industry leaders can retaliate, possibly to the extent of completely driving away or even bankrupting a new entrant. Because of their strong position, they could cut prices temporarily and destroy the whole financial viability of the new competitor. They could, if appropriate, engage in a major advertising or promotional campaign which nullified the more modest approach of the new entrant. They might be able to launch a new service that undermined the very basis on which the new competitor was hoping to build.

What, then, are the tactics which we should use to attack an industry leader? First, we should realize that industry leaders may be vulnerable under certain circumstances, and take steps to exploit this vulnerability. They can become complacent and even arrogant, feeling that they have a right to leadership for ever. They dismiss the early moves of competitors as insignificant. There was a world leader in a particular field of specialist software who began to notice some competition from a distributor in the USA whom they dismissed as a 'cowboy' of no consequence. They allowed him to gain 5 per cent, 10 per cent and then 15 per cent of the market, realizing too late that he had become a major threat. Being a low-overhead company, the competitor had been able to buy market share at prices that the multinational could not match. Having established his position, he was then able to raise his prices and continue to run a very profitable business. The approach used successfully in this true case may well be open to others.

In many fields of high technology services, such as management consultants, trainers and professional experts, competitors can be relatively small and can operate without the massive expenses, organizations, systems and other overheads of the larger players. If we are in this position, we have an enviable business advantage. We have tremendous freedom compared with the large operators, to price low and take the volume or to price high and take the margin.

Second, we should employ outflanking tactics rather than attacking head on. We should choose the battlefield, rather than fighting on the leader's own ground. If it comes to retaliation, the leader can always win. This might involve, for example, starting with a part of the marketplace or range of services with a lower profile than some of the more exciting areas.

Third, we must have a competitive advantage – some form of differentiation which is valued by the marketplace. We need to find some way in which we become a preferred supplier for at least part of the industry leader's business. Price is the differentiator that most readily comes to mind but, as we suggest in Chapter 6, this is the last differentiator we should use. By presenting the benefits rather than the features and by targeting those benefits (Sections 2.4 to 2.6), by creating uniqueness and differentiation (Section 2.7), and by creative use of segmenting, targeting and niche marketing (Chapter 4) we should be able to avoid the trap of the downward price spiral.

Fourth, we must be able to sustain the attack. It is worse than useless to start a battle that we are not able to continue.

Finally, the most successful strategy for attacking the industry leader, if the opportunity arises, is to choose a time when the leader has a serious problem. This can occur, for example, through an event which causes adverse publicity, a failed development project, the loss of key staff, or some other occurrence which diverts management's attention and temporarily lowers the barriers to entry. While we cannot create such an opportunity, it is worth having a strategy 'on the shelf' to bring into effect at the right moment.

Many readers will themselves be working for industry leaders. In this case they should interpret the above remarks in the sense of defending their leadership position; the tactics described are the ones which are likely to be adopted by their competitors. They should be on the lookout, and be prepared to repel serious-looking new entrants before they can make too much headway.

EXERCISES

5.1 Write down an example from your own company of as many as possible of the five types of competition in Section 5.1. What arguments would you use for winning a contract in each situation?

5.2 Analyse your company's competitors using the nine headings in Section 5.2 (adding others if you can). What competitive problems or opportunities does this analysis reveal? Set up a system for maintaining this information on an continuing basis.

FOLLOW-UP

5.1 Go round the organization asking the relevant staff what use they are making of the seven sources of information on competition listed in Section 5.3. Take extra time during the next six months to monitor these. What new strategies are your competitors adopting, and what counter-measures do you need to take?

In the light of this information, propose some specific ways in which this competitive information could be harnessed more effectively. Circulate a brief report on your findings to those involved, and arrange a meeting to decide what action should be taken as a result.

6

HOW DO WE SET PRICES?
AN UNDER-EXPLOITED MANAGEMENT OPPORTUNITY

Key business issues	Section
O Many companies under-price and some over-price. Pricing deserves far more management attention than it normally receives	6.1
O Value pricing needs to be understood by all who are involved in setting prices	
O Pricing is often too mechanistic, and financial departments have too much influence compared with marketing	
O Prices should be based wherever possible on what the market will bear rather than the cost of providing the service	
O New attitudes to pricing may be required, particularly when companies move outside their traditional field	
O In a free market the customer does not know or care about our cost (with some exceptions)	6.2

○ Different parts of the business have different profit
   potential   6.3

○ 'True cost' is impossible to calculate because it contains
   arbitrary allocations of indirect expenses   6.4

○ Marketing should do everything possible to avoid a
   commodity price situation, where all services are
   perceived as equivalent and people buy the cheapest   6.5

○ Although the price the market will bear is imprecise,
   there are several factors which can give an indication   6.6

○ Pricing should be used both strategically and tactically   6.7

○ The price of a service is rightly or wrongly taken to imply
   something about its value

6.1 VALUE PRICING – PRICE IS WHAT THE MARKET WILL BEAR

Pricing is one of the most under-exploited management opportunities. Experience with a large number of companies shows that many of them set their prices too low while others set them too high, because they do not understand value pricing. In the first case they are losing margin; in the second, they are losing volume. In both they are losing profit. The art is to use pricing to maximize total profitability.

In Section 1.10 we referred to three components of profit – market share, market size and percentage margin. We made the point that pricing is a fundamental part of the marketing function. It cannot be carried out in financial departments without reference to the marketplace or the competition.

What do we first want to know when setting a price? Most people would say 'the cost of producing the service'. This is *not* the right place to start. (To be consistent, we use the word 'price' throughout for what the customer pays, and 'cost' for what it costs us to deliver the service.) We should begin by asking 'what is the price which the market will bear?' Of course we need to know the cost, and ultimately we have to decide whether or not we want a particular piece of business, but the cost of providing a service is not the main factor determining what the customer will pay. At this stage, we should deliberately try to shut out from our mind any knowledge we have about our costs. The discipline of thinking first about the market and only then about our cost can lead to some very profitable conclusions.

In some circumstances, such as where we are providing people on a daily rate basis, the cost of providing the service is of course an important factor. In some cases, such as government tenders, the cost may actually have to be revealed. However, in most normal free markets, the price which the market is willing to pay and the cost of delivering the service are much less closely linked than people think. We do not have any automatic obligation to reveal our costs to the client, and should not do so unless we believe we have no choice. This causes a particular problem when companies move from one sphere of activity to another, perhaps as a result of diversification from their historic core business. They make the mistake of assuming that pricing decisions are made in the new business in the same way as they were in the old, because the management team has never seen it working any other way. This is simply not true, and can lead to some very costly errors. We would go so far as to say that, if they do not learn to adapt their thinking, they may never succeed in the new marketplace.

This works both ways. High costs, which were (presumably) necessary in the former situation and which were passed on under the old cost-based pricing regime, are totally unacceptable in the new circumstances; the customer is simply not willing to fund unnecessary activities, whether in the form of over-engineering, cumbersome bureaucracy or an inappropriate allocation of excessive overheads. In contrast to this, managers may simply not be aware of the higher margins which are possible in the new free market situation, because they are steeped in attitudes which were inculcated in an entirely different business environment.

Much of industry is wedded to the 'cost-plus' pricing philosophy. The cost is worked out, often to several significant figures. A certain percentage is then added on, in order to achieve the desired percentage margin or return on investment. The work goes on in the accounts department, and is carried out by people who have little or no contact with the marketplace or with the competition. This mechanistic approach to pricing can be disastrous in a free market.

It is impossible to achieve the optimum market-based price without considering the marketing issues which we have discussed earlier, particularly the benefits offered to the purchaser and the differentiation and uniqueness of the service being sold. A true case that dramatically illustrates this point is of a software engineer who used to be charged out by his company at £800 a day. He then changed employers, and his new company charged him out at £400 a day. How could they both be right? Of course they were not. The first company presented the value of this person and the excellence of the software skills he was providing in such a way that customers believed that they were getting a very good service (which they were). The second company set their sights so low in terms of image that they could not substantiate a higher price. The irony is that customers probably thought they were getting higher quality from the first company – 'if it's that expensive, it must be good' – even though the individual providing the service was exactly the same.

How can we find out what price the market will bear? The answer is – 'with great difficulty, and a large amount of uncertainty!' Some pointers to this are given in Section 6.6. In spite of the uncertainty associated with market-based pricing, one thing is clear; if we do not even try to find out, we shall make less profit than we should.

I have even heard as a justification for cost-based pricing that people are comfortable with it because they believe they can measure it accurately. Market-based pricing is imprecise, and therefore unacceptable to them. People who want neat answers to all business questions should keep away from marketing!

There is another key question which influences pricing decisions. How important to us is it that we win a particular contract? Do we need it desperately because otherwise our consultants will have insufficient work? Is it strategically important because it helps us to develop a track record in a new area? If so, we may choose to price at the lower end of the range of uncertainty in order to increase our chances. Alternatively, are we already so busy that further work will lead to capacity problems and excessive overtime, or to increased stress if we are a small operation? If so, we will probably price at the upper end of the range of uncertainty – if winning a contract is going to cause problems, we might at least make it worth while. With a cost-based approach to pricing, we would not reach these conclusions – we would probably not even consider the issues.

We are not suggesting that we should exploit customers by extorting an unreasonably high price from them. The discipline of basing price on value to the customer should actually prevent this, and we shall argue that some companies are over-pricing. We should separate the issues of price and cost much more clearly than is often the case. This is particularly true when we are moving from our traditional field to a new one, where the potential for profit may be very different.

If we can create good value for the customer at a low cost to ourselves, we deserve to make a good profit; if we cannot, we should not expect the customer to come to our rescue. A well-chosen price is a necessary part of the 'win-win' situation which is the characteristic of a good contract (Exercise 6.1).

To summarize, we should adopt the following mental approach to setting a price:

1.   Judge what price we think the market will bear in this particular case, by thinking about the value to the customer, without any reference to the cost.

2. Estimate the cost of providing the service.
3. Ask ourselves how important this business is to us, and consider whether we can reduce the cost.
4. Decide the optimum price.
5. Negotiate on the basis of value to the customer.

Three reasons are suggested for asking what price the market will bear before we think about the cost. They are discussed below.

6.2 CUSTOMERS DO NOT KNOW OR CARE ABOUT OUR COST

This is a 'black and white' statement designed to provoke! However, it is made because in free markets it is usually true.

Consider two suppliers, X and Y, offering equivalent services. The costs and selling prices (in £, £'000 or £'million – whichever makes the reader feel more at home) are shown in Table 6.1.

A customer asks the price from supplier X and is told '£110'. 'Ah, but I can get it down the road from Y for £100', is the reply.

At this point the customer is given the argument for paying £10 more. 'You see, my costs are high because I have a rather inefficient operation, I have high overheads and I have to pay an even higher corporate overhead levy, and my staff are so busy with internal meetings, e-mails and procedures that their productivity is low, so I have to charge more!' This is obviously ludicrous, but is it any more ludicrous than some of the reasons I have heard for charging a high price? These include:

1. *The allocation of indirect expenses.* I know of one company where two different parts of the organization can quote for the same project management task. For historic reasons, one profit centre is allocated a corporate overhead, the other is not. This difference has traditionally been reflected in the prices quoted, in some cases to the same potential client! Not surprisingly, customers are not interested

Table 6.1		
	X	*Y*
Cost	70	60
Price	110	100

in the way in which 'bean-counters' artificially move indirect expenses round the accounts.

2.   *A move to an environment where the existing level of overheads cannot be justified.* Companies that have been offering a service where price was not the main issue (e.g. where safety criteria demanded an extremely high level of expense) are now wanting to offer their under-utilized resources in the general commercial world. They have the necessary skills and equipment – the only problem is their historic cost. Before any progress can be made, the senior management must agree that normal commercial rates can be charged. They would obviously like to be able to recover their overheads at the historic rate, but there is not much point in having a high percentage of nothing while plant and people lie idle.

They will almost certainly have to take some painful steps to reduce the overheads, but they may not be able to wait until they have achieved these savings before pricing at the lower market rates.

3.   *An internal royalty or equivalent.* Subsidiaries of some companies have to carry a royalty charge if they use the company brand name. The customer may well be prepared to pay a premium because of the enhanced value which the brand name implies, but will not pay it simply because someone in corporate headquarters says so!

There is no reason why customers should be expected to pay the premium in any of these cases, and yet in a cost-based pricing regime they are in effect being asked to do so.

Returning to our example, the sale naturally goes to supplier Y. Y's volume goes up, and the unit costs go down. The reverse happens to supplier X (see Table 6.2).

Supplier X is now faced with a problem. His unit cost is going up, and he is a cost-based pricer. He therefore increases his price from £110 to, say, £120. Clearly he is in a vicious spiral, and he will soon be out of business.

Table 6.2		
	X	Y
Volume	Down	Up
Cost	Up from £70 to £80	Down from £60 to £55
Price	?	?

The alternative is for him to remain in business by charging the market price in order to buy some time to achieve a healthier cost base. He may have to reduce his staff numbers, accept a lower dividend or profit share, move to a cheaper location or do something to reduce cost; this may be unpalatable, but he has to find some way of breaking the vicious spiral.

Now consider supplier Y. Sales are going up, unit costs are coming down, and the only competitor has introduced a price increase. If Y is a cost-based pricer, he will bring his prices down – but why on earth should he? There is one possible reason for doing so – his only competitor is clearly in trouble and he may choose to accelerate the collapse. Apart from this, Y is in a virtuous spiral in which he is free to put his prices *up*, or to choose the part of the price-volume curve that maximizes profit. A cost-based approach to pricing would not have led him to that conclusion.

The above example is obviously trivial and yet, in a more subtle form, these occurrences do actually take place in well-known companies. It is not usually as overt as this. The time for a price review arrives. A hurried meeting takes place, and the prices are amended. In these circumstances there is no time for a considered review of market-based pricing factors; all we can do is to see how the costs have changed, and reprice accordingly.

It must be said that there are two areas where customers may care about our cost. If they think we are making too much profit, they will not be willing to pay our price. At the other end of the scale, if we are a regular supplier, a large purchaser will not want to bargain so hard that they drive us out of business. However, this usually leaves a wide range of possible prices between the two extremes.

Also, there are some cases where the customer is the owner of the supplier's business and has an absolute right to know the costs, but this is not a usual occurrence in a free market.

With these provisos, the first fallacy of cost-based pricing is that the customer does not ultimately know or care about our cost (Exercise 6.2).

6.3 DIFFERENT PARTS OF THE BUSINESS HAVE DIFFERENT PROFIT POTENTIAL

The second fallacy of cost-based pricing is that it appears to assume that there is some rule of business that enables us to make the same percentage profit on everything we do. This leads to the adoption of certain norms that become built into the pricing decision-making process.

This assumption is entirely without foundation. A balanced portfolio of high-margin/low-volume and low-margin/high-volume activities is perfectly normal. The aim is surely to maximize the profit of our entire business activity, not the percentage margin of any particular sub-component.

It is quite legitimate to say that there is a *minimum* percentage margin and a *minimum* return on investment that we are prepared to accept, and below this we will not engage in the business. This is quite different from saying that we will use fixed or minimum percentages as the basis for pricing. Again, failure to realize this leads to over-pricing in some situations and under-pricing in others, with the inevitable loss of profit.

This is another case where extrapolating from one business environment to another is dangerous. I can recall being a member of a new management team that began to question the cost-based price structure of an important product range which we had inherited. We found, after spending some time considering the market price, that we were under-pricing by a factor of two! How many other examples of incorrect pricing could be found if the necessary management time could be devoted to it (Exercise 6.3)?

These issues are so important that they lead us to conclude that many management teams should spend substantially more time on pricing. The resulting profit improvement could be surprising. Price increases that raised the average margin by as little as 1 per cent would have an enormous effect on return on investment in most companies. If one of the principal tasks of senior management is to increase return on investment, why do they not spend much more time wrestling with the issue of market-based pricing? (Follow-up 6.1.)

6.4   THERE IS NO SUCH THING AS 'TRUE COST'

There is a third factor that makes cost-based pricing not only inappropriate but impossible. It is that there is no such thing as 'true cost' anyway.

This statement may come as a surprise to people who live in a world of computer printouts giving minutely detailed costs. It is certainly possible to allocate some expenses fairly accurately when they can be directly attributed to a particular activity.

However, when we come to the 'true indirect expenses', such as strategic marketing, general and administration expenses, senior management expenses, site costs and so on, the allocation is inevitably going to be arbitrary. The normal method is to allocate them as a percentage of turnover, but is this necessarily 'real'? Why not allocate them on the basis of the added value that each activity generates, or the amount of time the managers spend thinking about it, which would give a radically different answer?

When discussing this issue, one managing director claimed that the allocation of overheads was 'very difficult but we do it very precisely'. Surely the truth is the opposite – it is relatively easy, but the result has a high degree of arbitrariness. One organization, subject to scrutiny by an official external auditor, found that there was a range of 30 per cent uncertainty in the 'true cost'.

This is not to say that there is no value in 'activity-based costing'. It is important to know where expense is being incurred. The problem is that we can only accurately allocate a proportion of the total cost. To use this method for determining prices is therefore flawed, quite apart from the fact that it is the wrong way of doing it (Exercise 6.4).

6.5   THE DANGER OF A COMMODITY PRICE ORIENTATION

The word 'commodity' in this context implies that there are a number of services that, although not necessarily identical, are broadly equivalent or substitutable. Typical commodities, listed in the financial sections of the daily papers, are coffee and nickel. Their price follows the well-known principle of supply and demand; when supply exceeds demand, prices fall and vice versa.

In these circumstances buyers will hunt around until they find the cheapest. In other words, in a commodity price situation, price is the most important factor in the buying decision. It is top of the list in the marketing mix, and the other factors hardly play any part as long as the specification is met.

There is nothing wrong with commodity trading – it has its rightful place in the commercial world – but it does not follow the normal rules of marketing and it has little to do with the marketing of high technology services. The danger is that people assume that they are in a commodity market when they are not. It is simply not true that people always use the cheapest consultant, accountant or lawyer, or always buy the cheapest company car, restaurant meal, computer or office furnishings.

The commodity approach is the opposite of the marketing philosophy that relates price to value. Part of the job of marketing is to bring price lower down the list of priorities in the mind of the buyer. By pushing the benefits to the top of the list, we attempt to reduce the significance of price.

Markets tend to move towards a commodity position as they mature. When a service is new, it is relatively easy to give it a meaningful differentiation. In a mature market, many services exist and they tend to be very similar; differentiation and the price premium that this can bring are much more difficult to achieve. The rate at which this takes place is accelerating in many fields as life cycles become shorter. Today's unique selling proposition is tomorrow's commodity.

A commodity orientation is death to professional marketing. A sophisticated high technology company cannot afford to sell at rock bottom price. If we cannot prevent such an occurrence, we might as well get rid of the marketing department, have very cheap premises, minimal overheads, negligible infrastructure and no investment in research. These actions are not compatible with a high technology company seeking to develop world-class services.

One common practice that actually encourages clients to think in terms of commodities is to quote hourly or daily rates. This suggests that we are selling our time rather than customer value. There is the silly story of the man who called roadside assistance when his car broke down. The mechanic arrived, took one look, went to his van to fetch a hammer. He struck the engine one hard blow, whereupon it began to run sweetly. He presented his bill for £50. Perturbed by this figure for such a simple job, the motorist asked for an itemized invoice! The mechanic wrote him one: 'for hitting engine with hammer, £5; for knowing where to hit, £45'! On a more serious level, a well-known computer company received a call from an overseas customer saying that their mainframe computer had crashed, they had lost all their data, and the company had ground to a halt. Could anything be done? 'Fly across with the hard disk stack', the computer company said. They did so, and within two hours the data was fully recovered. Then came the question of the invoice. The bean-counters looked at the records, and found that two person-hours had been used; at £75 per hour, the total fee was £150. The customers were delighted. How much would they have been willing to pay? Twice that? Certainly. Five times that? Probably. And they would not have felt that they were being cheated or exploited. The point is that the computer company was not selling its hours; it was selling thousands of years of accumulated experience. The fact that it could quickly diagnose and cure a problem *increased* the value of what it was doing – not decreased it. We should not under-sell ourselves. It is much better to quote a price for a project, with agreed performance criteria, than to fight the battle of the hourly rate. The fact that we have normally declared these rates in the past does not automatically mean that we should do so in the future.

One of the tasks of marketing is therefore to keep taking measures to lift the service out of the commodity orientation, by introducing new benefits and enhanced value in the mind of the customer. By first creating and then communicating a higher value in the service they are offering, they can command a higher price. We might argue that this is one way in which the marketing function can help to justify the cost of its existence. Beware a commodity price orientation (Exercise 6.5)!

6.6   HOW DO WE DETERMINE THE MARKET PRICE?

We have said that determining the price the market will bear is not always easy. Nor, to be fair, is there one single 'right price'. We can and should take steps to find out the general price range which the market is willing to pay, but, in the end, it is what price an individual customer is willing to pay on that occasion that matters.

The general phenomenon is what is known as 'price elasticity of demand'.

This attempts to quantify the amount by which sales will increase if we reduce the price by a certain amount and vice versa, and is typically used in a high volume sales situation (see Section 6.7).

For businesses where the unit sales volumes are much lower, and every contract is different, the statistical approach is not practicable and other means have to be adopted. These can include the following alternative approaches.

TRYING TO ASSESS VALUE TO THE CUSTOMER

In some cases value to the customer can be calculated reasonably accurately. If the result of our services is to increase the productivity by x per cent, or reduce the amount of waste or fuel consumption by y per cent, we can actually sit down with the client and do the arithmetic. We still have to decide what percentage of the savings or increased profit we can claim as our fee, but at least we have a rational basis for discussion. In these circumstances, some consultants would be prepared to be creative in their pricing strategy. They might ask a two-component fee, part fixed and part depending on the results achieved; at its extreme, they might be prepared to quote 'no cure, no pay'.

The problem comes when it is simply not possible to calculate the return on investing in our services. I know because this is true of my in-house seminars. I certainly try to allude to this, by estimating how much value would be created by winning one extra client or by increasing a price as a result of my teaching. One can certainly arrive at an order of magnitude that can be used in the price negotiations. The fact that we cannot quantify the value precisely does not mean that we ignore the approach; on the contrary, we have to work that much harder.

TRY TO FIND OUT COMPETITORS' PRICES

This is a necessary part of a mature approach to marketing, but there is a trap we must avoid. Some people find out the fees charged by one or two competitors, and then say that that is the price they will charge.

In many fields of consultancy there is an enormous range of fees between the top and the bottom. I have a list in front of me where the daily rates at the top of the range (a partner in a major international consultancy) are *ten times* those at the bottom (a technician in a low-cost contractor). Obviously we are not comparing like with like, but people tend to pitch their fees lower than they need because of some competitor who is claiming to offer a similar service at a lower price. Even when fairly similar levels of sophistication are compared, there is a difference of a factor of two or three between the top and bottom of the range. We need to have sound market-based arguments for the point within, above or below the range of competitors' prices where we pitch our own prices. If we have confidence in the incremental value we are offering, we should expect to be able to persuade customers to pay for it. If we do not, we had better look at the whole basis of our business (Exercise 6.6).

TRY TO FIND OUT THE AVAILABLE BUDGET

An innocent question such as 'is there a budget for this?' or a more direct question 'how much is in the budget for this?' may well elicit a response; (obviously we have to judge whether the response is true, or is simply a negotiating tactic). If the budget is £50 000, there is no point in quoting £51 000 unless we are prepared for what may be long drawn-out negotiations at a more senior level.

SUGGEST A 'GUIDE PRICE'

We could try to test the situation without doing anything irretrievable. We might quote 'a figure for budget purposes' and see what reaction we receive. If it becomes clear that we have pitched the price wrongly, we may be able to modify it (more easily downwards than upwards), perhaps using a changed specification as the justification.

TRY TO FIND OUT AUTHORITY LEVELS

In most organizations, individuals and committees have certain levels of authority within which they can spend their budget without reference to more senior management. This is perhaps a more sensitive issue than the budget level and it should be approached with care, but a question such as 'if it were less than £10 000 would you be able to approve it?' might elicit a response.

POSE ALTERNATIVE SPECIFICATIONS

If we are able to offer alternative specifications, such as the basic or an enhanced version of our service, we may discover how much the buyer is prepared to spend and how they are thinking about the value of our service.

A DIRECT QUESTION

Direct questioning might take the form 'how much are you prepared to pay?' I have taken the rather unusual approach of asking a group of seminar delegates how many of them would not have come if the price had been higher by £50, £100 or £150. While not pretending that this was an accurate measure of price elasticity, it certainly produced some useful information!

The above approaches should not be used in an underhand way, and we will only be successful in the long run if we are offering good value to the marketplace. Persuading customers to pay too high a price is not the basis for a good business, quite apart from the moral issues it raises.

Like democracy, trying to find out market prices has its limitations but any alternative is far worse. The better we are at doing it, the higher our profitability is likely to be (Exercise 6.7).

6.7 PRICING SHOULD BE USED BOTH STRATEGICALLY AND TACTICALLY

So far, we have argued that price is a key part of the marketing operation. We have urged that the benefits of the service should be presented in such a way that price is not the main factor in the buying decision. This philosophy should be built into our pricing strategy.

However, there are times when tactical use of pricing is appropriate. When launching a new service, or embarking on a new marketplace such as a different market segment or geographical region, the first few sales are absolutely crucial. If we or our service are unknown, the lack of profile actually constitutes a barrier to purchasing. It would almost be worth giving the service away in order to begin to establish a track record! More seriously, special offers or discounts may be worth while, but we must be careful not to undermine the perceived value of what we are selling. It might be better to announce the full price to establish the service's positioning, but offer concessions to obtain the initial orders. There may be times when it would be worth offering a small amount of 'free consultancy' in order to establish credibility and begin to build a relationship; this may be preferable to offering a cheap price.

THE PRICE–VOLUME RELATIONSHIP

An orthodox shape of the price–volume curve is shown in Figure 6.1. It is a theoretical model used by economists, and makes certain assumptions about the marketplace and the competition, which do not always apply in reality. Nevertheless, the model is a useful starting point in our thinking about pricing strategy and tactics.

Our turnover, and therefore our profitability, vary at different points along the price–volume curve. We have an infinite choice of options between high margin/low volume and low margin/high volume. The profitability calculation is, of course, affected by the fact that there is almost certainly a cost–volume

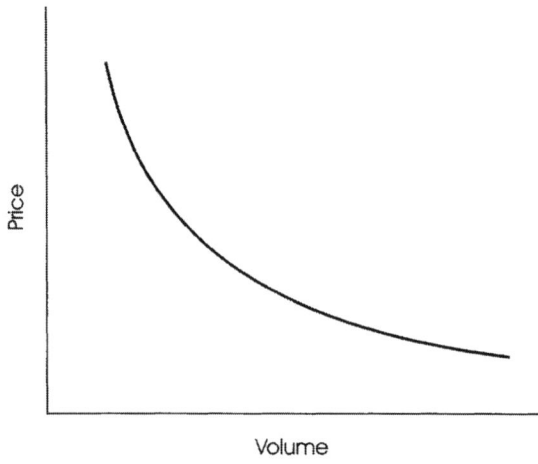

Figure 6.1   The price–volume relationship

curve as well; higher volumes achieved as a result of lower prices can reduce our unit costs and thereby generate a further contribution to profit.

The curve in Figure 6.1 is referred to as 'price elasticity of demand', and is defined as:

$$\frac{\% \text{ change in demand}}{\% \text{ change in price}}$$

For example, if reducing the price by 1 per cent causes the demand to increase by 2 per cent, the elasticity is 2. If we have to reduce the price by 4 per cent to achieve a 2 per cent increase in demand, the elasticity is 0.5. When the figure is more than 1, the demand is said to be elastic; below 1, it is inelastic. (Strictly speaking, the figure is a negative quantity, but this is usually ignored.)

For fast-moving consumer goods, such as soap powder or baked beans, where tens or hundreds of millions of units are sold each year, and each buying decision is similar to all the others, the suppliers will test one price against another, perhaps in separate but comparable locations. Their aim is to determine the slope of the tangent of the curve at the price level chosen – i.e. how many more or fewer units they could expect to sell if the price were decreased or increased by 1p, 2p, 5p and so on. With a high technology service, every buying decision is different, and we cannot draw smooth curves; nevertheless, the basic philosophy is identical and should be considered.

There is a problem, of course, in that the buying decision is a multi-factor operation. The rather simplistic curve assumes that all other things are equal

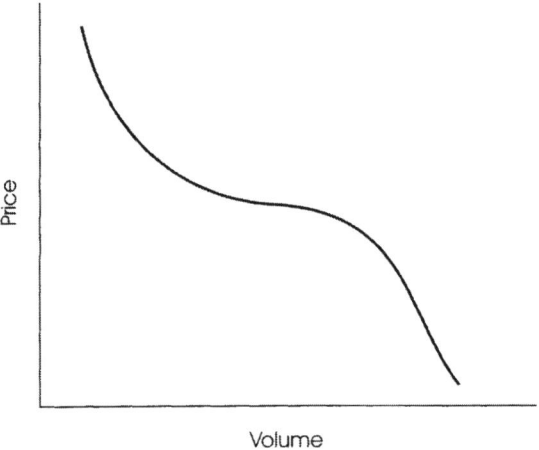

Figure 6.2 A saturated market

– which they never are. Competitors and customers have a habit of doing unpredictable things! Nevertheless, this approach can help us to clear our thinking and force us to ask questions which we might not otherwise ask.

However, we must offer a word of warning. The curve may actually take a quite different form. A second possibility is shown in Figure 6.2.

The left-hand part of the curve, representing the higher price region, is reasonably orthodox; but something strange happens as prices come down. At the lower end, however much we reduce prices, the volume does not appreciably increase. This is because the market is saturated. In this situation, reducing our price still further is the worst thing we could do – we would lose margin but not create extra demand. Somehow we have to find a way of reaching the orthodox part of the curve where we can maximize profit in a more healthy price–volume relationship.

A similar situation arises when new suppliers try to enter an existing market that is already crowded. Prices are already depressed, and coming in at a lower price still is unlikely to be profitable. Somehow they have to justify a higher price by meaningful differentiation.

A third possible price–volume relationship occurs when price represents a strong segmentation criterion. Figure 6.3 is a diagrammatic representation of three market segments, each of which follows a reasonably orthodox pattern.

An example might be computer software. At the top of the market, sophis-ticated tailor-made software is written for clients whose needs are so specia-lized and whose requirements are so demanding that nothing less will suffice; price is not the main consideration. The second section might represent the basic software for a personal computer, with good technical support and

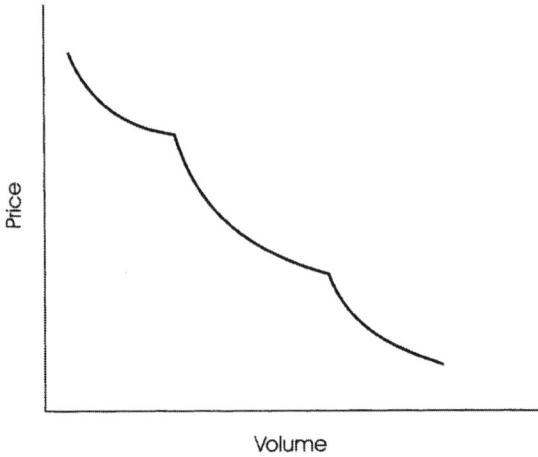

Figure 6.3   A segmented market

frequent updates; prices are generally reasonable, considering the huge range of capabilities offered. The third and lowest part of the curve is the cut-throat area, with standard 'shrink-wrapped packages' sold by 'box-shifters', where price overrides all other considerations and where customer support may be minimal.

An interesting strategic question is whether a company can operate credibly in more than one segment. There is a danger of confused brand image, so that the more expensive (and more profitable) offerings are undermined by the cheaper. It might be better to use different brand names or even different companies for targeting the different segments.

A rather surprising but very important price–volume relationship is shown in Figure 6.4.

At the top of the curve the volume increases as the price falls, but lower down the curve the volume actually *decreases* with a further drop in price. This illustrates the important point that price provides a signal about the value of the service we are selling. If something is too cheap, the perception is that it cannot be any good. This applies to hardware products, but it applies even more to intangible services where it is harder to attribute a 'correct' price. If consultants, scientists or engineers are much cheaper than average, we might think very carefully before entrusting the future of our business to their care.

STRATEGIC AND TACTICAL PRICING

Several options are open to us on strategic and tactical pricing.

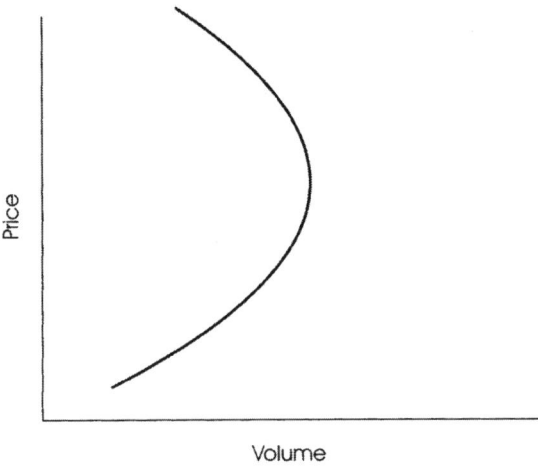

Figure 6.4 The effect of price on perceived value

Skimming

Skimming is when we price at the top of the market, as we might skim cream off a jug of milk. We are not attempting to target the mass market – we make our profit from a high-margin/low-volume approach, leaving competitors to do the opposite.

We have to convince the customer of the value of what we are selling. Curiously, with skimming, the high price may actually constitute part of the perceived value – people want to be seen to be working with prestigious consultancies.

Premium pricing

The premium pricing approach would be adopted where people are prepared to pay for excellent quality, sophisticated services and so on. The suppliers use the price–volume curve to their advantage. This presupposes that they are sufficiently familiar with the marketplace to be able to predict, at least broadly, what the reaction to the chosen price will be. For many high technology service providers, offering a differentiated service at a premium price should be their normal approach.

Normal competitive approach

Normal competition occurs when a number of companies are selling broadly equivalent services in the same marketplace. As has been explained throughout this book, the important issue is to differentiate our service from the competition's and to do everything possible to bring price down the ranking of criteria determining the buying decision.

Rock bottom pricing

This approach would be used when the overriding consideration is to achieve volume; for example, when we have a very under-utilized resource where the fixed costs are having to be paid anyway. The argument is that it is better for us to sell services at a price that makes at least some contribution to fixed costs than to have no contribution at all. It also makes it almost impossible for competitors with lower volumes to compete at a profit.

The rock bottom approach might also be seen in the case of a very high-value tender. In these circumstances, there will almost certainly be a number of suppliers who are willing to offer rock bottom prices because they desperately need the volume.

While rock bottom pricing may have some role in a multi-service environment where other services are making good margins, or in a short-term exceptional occurrence, it can be highly dangerous if our whole operation is based upon it. It does not represent a basis for running a viable business. We have to find some way of moving up the price scale, by convincing the buyer that the value of what we are offering is greater than the others and therefore justifies a premium price.

This discussion again confirms the point that a great deal of senior management attention needs to be devoted to pricing. Obviously the various tactics are meaningless if we start with a cost-plus approach to the pricing operation (Exercise 6.8).

EXERCISES

6.1   Examine some recent pricing decisions made by your organization. To what extent were they based on 'what the market would bear' rather than 'cost plus'?

6.2   Do you think your customers know about your costs? Do they care? What can you do to promote the *value* of your service, in order to reduce the linkage between price and cost?

6.3   What is the range of percentage margins for your services? Is this range based on a good assessment of what the market will pay for the different services, or have the prices been influenced by some internal 'norms'? What opportunities for profit improvement does this offer?

6.4   By what percentage could you reasonably change the 'cost' of one of your main services by reallocating some fixed expenses? Would this make any difference to your pricing decisions?

6.5 Are any of your services 'commodities', or are any moving in that direction? What can be done to lift them out of the commodity environment, or at least to reverse the trend?

6.6 Rank your competitors' prices for equivalent services in a scale of descending order. How do you justify your position on, above or below this scale? Does this lead you to reconsider your approach to pricing?

6.7 What further steps can you take to find out the price the market will bear?

6.8 How accurately do you know the effect on volume of increasing or decreasing your prices by 1 per cent, 5 per cent, 10 per cent, 25 per cent etc.? Are you sure that you are pricing correctly? Is there one service for which you could do some experimenting with different pricing without too much risk in order to test the elasticity? If not, could you do some market research into the same area?

FOLLOW-UP

Over the next six months (or other suitable period, depending upon whether you have an annual or other cyclical pricing regime), record the time you spend on preparing for and making pricing decisions. *Exclude any time spent on costs.* If at all possible, ask your colleagues up to and including chief executive level to do the same. At the end of the period, ask yourself what would have been the effect of spending twice this amount of time. Do you think this would have been a profitable use of senior management time? (Do some arithmetic on the likely results.)

PART II
THE ROLE OF MARKETING COMMUNICATION

❖

7

HOW DO WE COMMUNICATE?
PRINCIPLES OF COMMUNICATION

Key business issues	*Section*
O Many different means of communication are available to us. Each has a different purpose and should be used accordingly	7.1
O A considerable amount of money is wasted because people do not ask 'What are we trying to say?', 'To whom?', 'With what objective?', 'Through what medium?'	
O Communication is a progressive process, starting from zero and leading to signing a contract. The purpose of each element of the communication is not to sell the service but to sell the next step	7.2
O The key to cost-effective communication is qualification of leads. We want quality of leads not quantity	
O Generalizations such as 'marketing does not work in our business' are often very superficial judgements based on an inadequate understanding of marketing	7.3

○ Extrapolation from one selling situation to another can be highly dangerous. Methods which work in one case may be inappropriate in another

○ The size and diversity of the target audience is a key determinant of the forms of communication to be used

○ Impersonal means of communication should be used in the early stages of the communications process to generate qualified leads. Expensive personal selling should then be used to follow up

○ External agencies are experts in communications. We still have to take responsibility for the message   7.4

7.1 THE COMMUNICATIONS PROCESS

The technical director of a research association once said on one of my seminars, having made sure that the marketing director and managing director were listening, 'there is no evidence that we have ever won a contract as a result of one of our glossy brochures, so we should stop wasting money on all this marketing nonsense and spend it on technical equipment instead'. Was he right? Was the money being spent on brochures indeed being wasted? This raises the whole question of the purpose of a brochure, or any other form of marketing communication.

In our own daily experience, both business and personal, we may well ask the same question. We are bombarded with a vast amount of marketing communication. It takes a variety of forms – advertisements, brochures, letters, telephone calls, faxes, visits and so on – and we may understandably wonder how effective it all is.

A large proportion of this communication is ineffective because some very basic questions have not been considered:

○   What are we trying to say?
○   To whom?
○   With what objective?
○   Through what medium?

WHAT ARE WE TRYING TO SAY?

When asked to engage in some form of marketing communication, the instinctive reaction of most people who have not been trained in marketing is to tell people about their service. They support their presentation with lists of specifications, functionalities, skills, past achievements and so on, the whole emphasis being 'this is what we do'. At first sight this sounds perfectly reasonable but it is, in fact, quite the wrong way of going about it as we shall explain in Section 7.2.

TO WHOM?

Communication is a two-way process – communicating *with* people rather than *at* them. It starts with finding out something about the potential customer and his or her needs. Then, and only then, can we go on to say what we have to offer because, by this stage, we have learned at least something about what they might need and we can target our message accordingly. This important aspect of the communication process is discussed in Section 8.2.

We often have to 'sell' to a number of different people, each of them with different perspectives and different needs (see DMG in Section 2.6). Until we are clear about the target at which a message is aimed, we cannot possibly compose the message itself.

WITH WHAT OBJECTIVE?

This is where the biggest mistake is made. The objective is to sell, of course. Wrong! This is only true when the sale is made during a very simple encounter, for example, in a shop. We can now consider the issue raised by the technical director at the beginning of this chapter. The objective of a glossy brochure is not to win a contract; it is to generate a phone call, a request for further information, or a request for a meeting – a single, positive but limited stage in the process. Therefore we can say that the objective of a single piece of marketing communication is normally to 'sell the next step', and that this is part of a multi-stage process (see Section 7.2). If it does this, it has achieved 100 per cent of its objective.

THROUGH WHAT MEDIUM?

The word 'media' normally refers to different forms of advertising – press, magazines and so on. However, it is useful to broaden this to include every form of marketing communication. We have a wide variety of media at our disposal – personal contact, literature, proposals, advertising, presentations, e-mails, web pages, letters, telephone calls, exhibitions, public relations and so on – with an infinite number of ways of combining them. In each case we need to ask 'is this the most appropriate medium?' (Exercise 7.1).

7.2 'SELLING THE NEXT STEP'

In most of the cases in which readers are involved, the selling process is much more complex than a single encounter. It involves a number of stages and the whole process may take days, weeks, months or even years. The objective of any form of marketing communication should therefore be to 'sell the next step' in the process. If the 'next step' is to complete the deal, well and good, but there may have to be several intermediate steps before this is possible.

Where a sale takes place over an extended period of time, a large number of people may be involved in the purchasing decision. Similarly, a number of people from the selling company may become involved – seller to buyer, scientist to scientist, engineer to engineer, quality manager to quality manager, financial manager to financial manager, chief executive to chief executive and so on.

The important point is that a contract negotiation is usually a process leading up to an event rather than a single event on its own. If the extended process is not managed properly, the event – when the buyer actually commits to purchasing – may never be reached.

What are the stages in this process?

Let us start by assuming the most difficult case. The buying company has never heard of us. They do not know that they need our service, and even if they did, they would not take the initiative in coming to us. They would not put our company on their 'long list', let alone their short list. This is a particular problem when we are diversifying from our historic core business into areas where we have no track record, as discussed in Section 4.8. Unless we take the initiative, nothing will happen; the selling process has stopped before it has started!

(Actually, this might not be the most difficult case. We might start not from a low profile but from a negative profile – for some reason the buyer is actually predisposed against us. This might come about, for example, because of some perceived problem with our service or company, justified or otherwise.)

Another situation might be that the buyers know in a general sort of way that we are in the right field, but do not have enough information to decide whether or not to approach us. Again, we have to approach them. This is why the whole culture of 'reactive selling' discussed in Chapter 1 completely fails. Unless we are proactive, i.e. we take the initiative to go out and make the first contact, nothing will happen.

The next stage, after the customer has actually realized that we may have something to offer, is a growing understanding of what we have to sell; this is accompanied by a growing understanding by us of what the customer might need.

The needs which a customer initially expresses may not be the real needs, and certainly not the total needs. Part of the job of the seller is to help the buyer to understand what is really needed. If we succeed in doing this, we may not be guaranteed the sale but we are almost certainly in 'pole position'.

The next stage is serious negotiation. In a simple case this will involve presenting our solution, answering questions, overcoming objections and progressing towards the close. In a more complex case, such as a tender for a large capital project, the process may involve proposals, presentations, meetings, visits by a number of people and so on.

This last stage is 'commitment to buy'. This takes place when the buyer has decided to proceed. In a large purchase, the decision may be 'subject to contract' (as when we buy a house). There may have to be important discussions over contractual, legal, logistical and other aspects; these are a crucial part of the selling process and the sale can be lost at this stage.

The point about recognising the progressive nature of the selling process is that, in many cases, only one of these stages can be achieved at a time. The art of effective selling is therefore to concentrate on the issues involved at each particular stage – selling the idea of a meeting, for example. The ultimate goal of achieving the order will of course be in the back of our mind, but it will not dominate our actions in the earlier stages (Follow-up 7.1).

'QUALIFIED LEADS'

If we send out a leaflet or put an advertisement in a journal, the initial probability of a sale might be one in a thousand or even less. If we speak on the telephone to someone who has received this first communication, the probability might increase to one in twenty. If we meet the person face to face, it might increase to one in two. The probability reaches 100 per cent when a contract is signed.

The point is that we are progressively 'qualifying' the lead. The objective is to obtain quality of leads and not quantity. It costs a great deal of money to process leads; contacts who are unlikely to buy should be eliminated or relegated to a lower priority as early as possible in the process, so that attention can be concentrated on those who might.

A CASE STUDY

The progressive nature of marketing communication can be illustrated by a true case study in which the author was involved. The objective was to sell the services of the high technology faculties of a university. They had had links with industry over at least 20 years, and continued to have some good customers. The objective, however, was to be pioneering and to gain business from companies with whom they had had no prior contact.

Various options were open to them. They might run a series of 'open days', to which invitations were widely issued. Alternatively, a number of technical experts might make forays into various parts of industry where they thought there might be an opportunity. What we actually did was neither of these, but it resulted from a careful analysis of the selling process as outlined above.

We started by making a list of potential targets by consulting appropriate directories. (An alternative, which would have saved time but which would have cost more money, would have been to purchase lists of companies in certain categories.) Targets were selected as being companies above a certain size within 100 miles and operating in technology areas where the university had expertise.

A 'cold mailshot' was then carefully written. A key part of the strategy was to understand that the objective of this mailshot was *to sell an interview*, not to sell the services of the university. A one-page leaflet was included to establish relevance and credibility, but the whole emphasis of the letter was that I should visit a senior person in the target company. The letter described the benefits of the interview as a means of linking the company with a wide range of skills without involving any commitment and without taking more than an hour of their time.

An important part of this approach was the 'reply form'. This is set out in detail in Section 9.4. Sixty per cent of the letters received a reply, and 50 per cent of these (30 per cent of the total) resulted in an interview on the customer's site, a very high percentage by most standards.

The second stage was to identify potential needs of the client. In some cases they knew of an existing problem which outside assistance might help to solve. In other cases, some judicious probing was necessary before they were able to identify a number of areas where outside assistance could be valuable. At this stage very little was said about the actual technology being sold. The whole emphasis was on providing business solutions, within agreed costs and time-scales. The objective was to achieve the third stage, a meeting on our site. The conversion ratio of visits to the client into structured meetings on our premises was about 75 per cent (22 per cent of the total), again a very high figure by most standards.

At these meetings, an agenda had been agreed as a result of the initial visit, and the right technical staff were present. They probed the needs more deeply and then described how they would address the particular problems that had been identified. Where appropriate, they demonstrated the facilities to the potential client, and described past successes in similar areas. The objective of the meeting was to see whether it would be worth submitting a proposal.

Thereafter, the process went in the normal manner of invitations to tender with a contract ultimately being awarded or not. The most important stage is the process that led up to this invitation to tender. After all, the client had not come to us. We had gone out to the client, assuming that there might be some business but with no actual demand expressed (Exercise 7.2).

Having established that the marketing communication process involves a number of very different stages, and readers will no doubt be able to amplify these, we can now examine the role of the various elements we are able to use.

The important point to recognize is that each element is there to achieve a different objective during the progressive communications process. The way in which each is used will depend on a variety of factors – the size of the target market, the value of a potential contract, the number of people involved in the decision, the length of time likely to be taken for the whole process from start to finish, and so on.

7.3 THE ROLE OF DIFFERENT METHODS OF COMMUNICATION

A qualified professional person once said to me 'We tried marketing – it doesn't work in our business'. Such a generalization simply showed that the speaker did not understand the marketing process. What she meant was that they had written a brochure that was lying in the basement gathering dust. Inappropriate and inept marketing does not work in any business, but we cannot dismiss the whole multibillion pound communications industry on such a superficial basis.

Another person, a highly qualified scientist, said 'When I get a journal, I open it over the wastepaper bin and tip all the inserts away; therefore inserts are a waste of money'. Let us suppose for a moment that he was right (although, interestingly, he was attending my seminar as a result of one of my inserts!). To extrapolate to the world from a sample of one shows a scientific inconsistency that would have caused him to fail his school examinations, let alone his university degree. We are dealing with statistical probabilities in this sort of marketing. We may all be nauseated at times by the 'junk mail' that arrives through the post, and obviously some of it is more effective than others, but the simple fact is that the advertisers do it because it works. They know that most of it will be ineffective, but they have done their arithmetic and they know that, if they achieve a certain number of replies, the campaign will not only have been successful but will also have been more cost-effective than other ways of achieving the same level of response.

The diagram in Figure 7.1 indicates the relative effectiveness of impersonal and personal forms of marketing communication at different stages in the

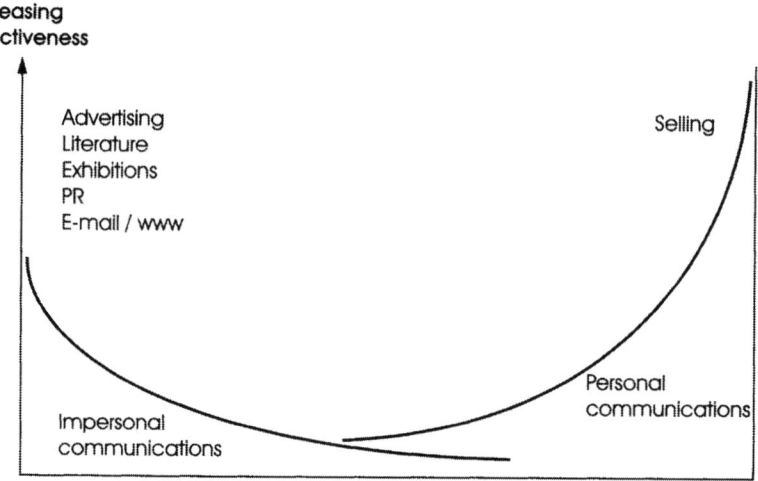

Figure 7.1 **The role of personal and impersonal communications**

selling process. The philosophy has been described in the previous section as 'selling the next step' and obtaining 'qualified leads'. The vertical axis is an unquantified measure of the relative effectiveness of each form of communication as the customer proceeds towards a purchase. The horizontal axis indicates the increasing awareness of the customer to the need for our service; the time-scale could be two minutes in a shop or two years in a contract negotiation – the principle is the same.

The effectiveness of *impersonal* media such as brochures and advertising declines as the communications process proceeds. Their role is to take the potential customer, at a relatively low unit cost, to the position where it begins to be worth employing *personal* selling efforts.

Quite apart from the cost of using personal selling early in the cycle, there is another very good reason why this should not normally be done. Sales staff are trained to close deals. They are highly motivated to obtain orders. In these circumstances, they are not going to be very excited if they are asked to give a lecture or travel around simply trying to generate interest or 'fly the flag'.

We must avoid making generalizations. The precise way in which impersonal and personal communications should be used depends critically upon the potential size and diversity of the target market. Figure 7.2 distinguishes four situations, A, B, C and D, each with a very different type of customer base.

Customer base	Route to market	Example of customers	Communications strategy
A Small defined	Direct to customer	Car manufacturers	*Personal* with whole market
			Impersonal (e.g. brochures, web sites, videos, case studies) to create image and support personal activities
B Large defined	Direct to customer	All major manufacturers	*Impersonal* narrowly targeted (e.g. direct mail) to qualify leads
			Personal with qualified leads
C Effectively infinite	Direct to customer	PC users	*Impersonal* broadly targeted (e.g. inserts, web sites, adverts) to qualify leads
			Personal with qualified leads
D Effectively infinite	Through sales channels	PC users	*Personal* with channels
			Impersonal (e.g. inserts, adverts, point-of-sale display) to create pull through and support channels

Figure 7.2   Size and diversity of target market

DIRECT APPROACH TO SMALL DEFINED CUSTOMER BASE

If we are selling services to car manufacturers, it would be ludicrous to send out a brochure and then decide that the manufacturer of millions of cars was not interested in our service simply because the brochure did not result in an order! The size of the target market is so small numerically that we can and must approach each one individually. The role of impersonal means of communication, such as brochures, videos, case studies and so on, is subsidiary; these create a general background image on the basis of which we can carry out our marketing strategies and support our personal activities.

DIRECT APPROACH TO LARGE DEFINED CUSTOMER BASE

Let us suppose we are selling a service that could apply to all large manufacturers. Examples might be computer integrated manufacture, quality management, standards conformance, training, safety, or other forms of manufacturing expertise. The target market is too large for the initial approach to be made personally in every case, even though we could 'prequalify' it by only including manufacturers above a certain size. The best strategy would probably be to use narrowly targeted impersonal communications, such as direct mail with a letter and brochure, in order to qualify the leads. Personal contact would then be made with those who had shown some response.

DIRECT APPROACH TO EFFECTIVELY INFINITE CUSTOMER BASE

By 'effectively infinite' we mean that the group is so large that there is no way in which we could personally contact the whole group. This is an important statement to make, because we may go on to conclude that a potential customer who has rejected our first approach may not be worth any further effort and we feel that we would be better to go on to new targets which have not already rejected us; the rejection represents a negatively qualified lead. (Against this, it could be argued that a second and third approach may reinforce the earlier one, so that we are more likely to achieve a response in due course. We have to make this judgement in the light of the particular circumstances.)

An example of an effectively infinite market approached directly would be all users of personal computers (PCs). We might use an advertisement or an insert in appropriate journals. The aim is to qualify the leads to determine which should receive our personal marketing efforts. We might follow up the leads from the advertisement with a personal letter as stage 2, and then try to 'sell a demonstration' to those who have responded to the letter as stage 3.

APPROACH THROUGH SALES CHANNELS TO AN EFFECTIVELY INFINITE CUSTOMER BASE

Again the customers might be users of personal computers, but the difference is that we are selling through wholesalers, retailers or some other form of sales channel. Our efforts would obviously be personal with each sales channel in terms of selecting, appointing, training and motivating them, but we would have no direct contact with the ultimate purchasers. Our impersonal marketing efforts would probably be aimed at creating 'pull-through'. This could involve providing the channels with some way of reinforcing their own sales efforts such as point-of-sale display material for their outlets and also, possibly, helping to fund their advertisements, inserts and so on. This expenditure would also encourage the management of the channels to make a corresponding effort.

By studying the four examples above, and relating our own business to one or other of them, we can see that the use of impersonal and personal means of communication has to be very clearly thought out. An enormous amount of money is wasted on untargeted impersonal communications, and our aim should be to increase the percentage response progressively as we learn from experience. Equally, the productivity of our much more expensive personal efforts can be increased significantly if we first qualify the leads and target our efforts accordingly (Exercise 7.3).

7.4 THE ROLE OF EXTERNAL AGENCIES

When asked about their marketing communications, some people reply 'Oh, our advertising agency does that for us'. An in-house publicity department might play a similar role.

We need to understand the different roles of the client and the agency or publicity department in the communications process. The reason we use expensive outside resources is that they are experts in *communication*, which most of us are not. However, the client has to take responsibility for most if not all of the *message*.

If a client has a very large budget for advertising, literature and so on, a good agency will dedicate people to serve that particular client. They will invest time in learning to understand the marketplace, the services, the customers and the competition. They will then be in a position to assist the client in developing marketing strategy and in articulating the message to be communicated, even though ultimate responsibility still rests with the client.

If we have a smaller budget, it is a matter of commercial common sense that an agency is not willing to devote such resources to the account. The responsibility for defining the message is almost entirely ours. If no communications expert is used, we have to do the whole job ourselves. This is not an unreasonable approach in many cases. The high technology expert is the best person to understand the needs of the client and to say how they can be met, and the capabilities of internal desktop publishing systems are very impressive these days.

Failure to understand the distinction between the message and the communication vehicle is the cause of an incredible amount of wasted money. It leads to companies producing very expensive glossy brochures that say very little (Exercise 7.4).

EXERCISES

7.1   Select a piece of promotional literature used by your organization, and a piece which you have received as a potential customer, and assess them against the four questions in Section 7.1.

7.2   Consider the case study in Section 7.2, and see whether a similar process could be used for promoting your organization into a new area.

7.3   Apply the framework of Figure 7.2 to your own situation. What ideas does it give you for improving your own use of impersonal and personal communications?

7.4   Consider the relationship between your company and external communications agencies. It is working well? Should you be doing more to create the message rather than leaving it to them?

FOLLOW-UP

7.1   Every time you are involved in selling, ask yourself 'what is the next step?' Concentrate on this rather than on the ultimate sale, until it becomes second nature to you.

8
HOW DO WE SELL?
PERSONAL COMMUNICATION

Key business issues	Section
O Selling can be learned. It is a fallacy to assume that salesmen are born	8.1
O Personality in a salesperson may be important, but far less than is commonly thought. Selling in a high technology business is an extremely rational process	
O Selling starts with listening rather than speaking. We should prepare open-ended questions which will progressively lead customers to reveal their real needs	8.2
O Presentations should be specifically prepared for each occasion and targeted to the particular customer. Standard presentations do not sufficiently exploit the opportunities	8.3
O It may be worth attending a 'presentations skills' course, but attention to the message is even more important	
O Objections should be anticipated and a reasoned answer prepared	8.4

○   There are many techniques for closing the sale. The
     sale should not be left to close itself   8.5

○   Selling can be carried out by full-time sales staff or by
     others (e.g. high technology fee-earners) as part of their
     normal work. There are advantages and disadvantages
     to both   8.6

○   A combination of sales and other staff can be the most
     effective, with generalists bringing in specialists at the
     appropriate time

○   Existing customers should be nurtured and used   8.7

○   A complaint that is well handled can actually increase
     loyalty

○   A good way for a small organization to conduct
     international business is by means of partnerships,
     but certain ingredients are necessary for success   8.8

8.1   SELLING CAN BE LEARNED

We often hear the statement that 'sales people are born and not made'. This
view is usually based on the caricature of a salesperson as a slick operator
with the 'gift of the gab' who earns high commission payments at the
expense of the unfortunate victims.

Readers will quite rightly reject this view as being totally out of character
for their own business. As high technologists, they have high professional
standards; the whole scientific method is based on honesty and objectivity,
and rejects manipulation and emotion. The good news is that selling, too, is
an extremely rational process which can be taught and learned. Self-assurance
and the ability to present and persuade are not enough, and do not make up
for the benefits that can be derived from a professional sales training
programme.

In a book that covers the whole of the sales and marketing function, it is
only possible to describe some of the main techniques of selling. If the
reader is stimulated by this introduction, it would be worth seeking out a
specialist training course, book or video.

8.2 SELLING STARTS WITH LISTENING RATHER THAN SPEAKING

'I keep six honest serving men,
(They taught me all I knew);
Their names are What? and Why? and When?,
And How? and Where? and Who?'

(Rudyard Kipling, 'I keep six honest serving men')

The instinct of many sales people is to bombard the hearer with a mass of facts and arguments that are supposed to lead to the placing of an order. Through ignorance or enthusiasm, or because the seller wants to keep control of the conversation, the whole direction of communication is one way – it is communicating *at* rather than communicating *with* the customer. It is telling what we do, not identifying what the customer wants.

We need to create a situation where customers start to talk about their business or personal situation, their problems and their needs. Only then can we supply a satisfactory solution. If this sequence is not followed, we may be advocating a remedy for the wrong disease or even a remedy for which there is no known disease! It is as if a doctor said: 'I'm so glad you've come to the surgery. You really must try this marvellous new drug. By the way, what's the matter with you?' A doctor first diagnoses, then prescribes; so should we.

Suitable questions should be posed at an early stage in the conversation, in a very unthreatening manner, in order to get the prospect talking. An over-aggressive approach will have the reverse effect.

When approaching a sales interview, it is worth preparing some questions in advance, which will begin to generate this two-way communication. Clearly we are not going to interview the buyer with an intrusive clipboard listing a number of questions to which we demand answers, but this does not excuse us from taking a professional approach to preparation. A useful ploy is to write down two or three questions on the second page of a pad of paper. The mere act of writing them down will probably register them in our memory but, if not, they can probably be seen through page 1, or the pad can be casually flipped open as a reminder. We must be prepared to depart completely from our prepared opening questions, but if we plan the interview there is at least a chance that it will go the way we had hoped.

These questions should be open ended and not capable of being answered 'yes' or 'no'. (You may need to ask some diagnostic questions first, but these should not be confused with the open-ended type of question designed to explore the buyer's real needs).

A useful approach is to ask the buyer what problems are being experienced, and then ask about the effect of these problems on the business. This problem and effect approach encourages the buyer to realize the extent of the need.

Even with this method, there will be times when the question does not lead to the desired response. It may be necessary to probe ('that's interesting – what exactly did you mean by that?'), to rephrase the question or to approach the subject from a slightly different direction; the objective is to encourage the customer to talk.

When we have begun to form a view as to the buyer's needs and motivations, and the 'hot buttons' which are likely to lead to a favourable response, then and only then can we begin to apply our solution (Exercise 8.1), (Follow-up 8.1).

8.3   PRESENTATIONS SHOULD BE PREPARED AND TARGETED

People rightly think that presentations should be carried out as professionally as possible. They prepare computerized multimedia presentations, handouts, overhead transparencies, slides, videos and other media which are aimed at presenting their products and company in the best possible light.

Laudable though this is, it can actually have a negative effect. The problem is that the mechanics of the presentation can get in the way of the message being presented.

Some presentation media are inflexible. If we have invested a great deal of time and money in a corporate video, for example, the instinct is to show it on every possible occasion. This is fine as long as we are presenting to homogeneous audiences whose needs are all the same. The problem is that they are not. One of my clients was good enough to confess that he had fallen into this trap. 'As I watched their reaction to the video' he said 'I could see their eyes glazing over – we were giving the wrong signals.'

The other extreme – not preparing anything – is equally ineffective. Experienced people feel that they have been doing presentations for many years and that 'it will be all right on the day'. They do not anticipate the questions and problems that concern that particular client, and therefore do not have the right material available. The client may well think that, if the company cannot even prepare a presentation properly, their product or service cannot be much good either.

The aim is to achieve a balance between these two extremes – preparing thoroughly but not allowing the mechanics to get in the way of the message. Some guidelines are given below.

1.   *Find out as much as possible about the client's needs in advance.*
        Selling starts with listening rather than speaking. If the presenter has

already met the client, this should not be a problem. However, if we arrange a 'presentation at head office' which is given by others, a clear brief must be given on the background of the client, the objective of the presentation and the desired outcome of the whole process. Too often visitors are given a standard tour and presentation, and the seller is left wondering why so few of the efforts appear to bear fruit.

2. *Decide the specific objective of the presentation.* Thinking back to Section 7.2, we have to decide what is the 'next step' in the communications process. What sort of commitment we are aiming to get from the client and how we are going to obtain it? It might be an agreement to submit a proposal, a further visit, a meeting, a discussion, a demonstration or a visit to a reference site. Whatever the objective is, it should be clearly understood by everyone involved in the presentation. It is more important to sell this next step than to do a hard sell on the product or service at too early a stage in the process.

3. *Be clear about the message.* Before doing the presentation, the person or team involved should write down in a few words the key points which they want to communicate. They should ensure that these points come across clearly and unequivocally throughout the presentation, in a way that can be remembered by the hearers. I like to apply the 'car park test'. When the members of the visiting client team are speaking to each other when they are first out of earshot of the presenters, how do they finish the sentence 'I think we should use these people because ...'?

4. *Concentrate on their needs not your skills.* This point is often particularly badly handled in presentations. It is much easier to talk about ourselves than about the client's needs. Success goes to those who realize this and take the trouble to overcome it.

5. *Be flexible.* In spite of everything which has been said above, we still need to be flexible on the day. Unforeseen questions and problems may arise, and it is no use sticking to our planned format if the client is temporarily concerned about something different. One way of overcoming this in a professional way is to have alternative presentation material available in a readily indexed and accessible form. As an example, I prepare my seminars to meet the client's brief as closely as I can. However, I normally have with me 20 copies of about 50 different handout pages which can be passed round immediately if

an unforeseen point arises. I use pages of different colours rather than numbering the pages, because I want to maintain flexibility. This balance between being very well prepared and being flexible is the key to success (Follow-up 8.2).

6.   *Attend a 'presentation skills' course if necessary.* Such courses are worth while but are not a substitute for the recommendations above. A client once said to me 'I attended a presentation skills course. At my next presentation, I avoided all the errors such as jangling keys in my pocket, and I felt that I had done a really good job. I told the audience that we were wonderful and we'd been in business for 50 years; unfortunately it achieved nothing'. It would be better to present a clearly targeted message in a slightly unprofessional way than to have a very smooth presentation of the wrong material. However, it is better to be right on both counts!

8.4   OVERCOMING OBJECTIONS

Objections are matters raised by the potential buyer, true or false, relevant or irrelevant, which stand in the way of a positive buying decision. Until we have dealt with a particular objection it may be impossible to make any further progress.

We will sometimes encounter objections we have not heard before, in which case we have to think very quickly and give the best answer we can. However, some of the same objections come up time and time again, and it is well worth preparing an answer in advance. These objections may include some of those below.

PRICE

Price is perhaps the most common objection. We could even argue that, if price is never an objection, we are under-pricing! As we said in Chapter 2, our aim must be to bring price down the ranking in the buying decision, by emphasizing the benefits that are most likely to be relevant.

The 'lifetime cost of ownership' is often more important than the initial price. The cheapest software package may be the worst possible purchase if the supplier cannot give the necessary support, upgrades and so on. A product or service which is cheap to buy but inefficient and expensive thereafter may be a very bad investment. The seller must present these arguments, even to the extent of carrying out an investment appraisal on behalf of the customer. This is particularly true if the idea has to be 'sold' internally to people who may look at the initial price and disregard the other financial implications.

UNCERTAINTY ABOUT THE IDENTITY OF THE SUPPLIER

If we are buying something like a car, we want to be assured that the manufacturer is reliable. In principle, we could do this by visiting their factory, examining their quality control procedures and so on, until we were satisfied. In practice, we do not do this. We look at their product, read the journals, and perhaps talk to people who already own that particular brand. We judge the supplier by the product.

In the case of an intangible service, we cannot do that. The product does not exist until the service is performed. Some world-class high technology services are supplied by relatively unknown suppliers.

To overcome this objection, we have to give evidence which turns the intangible into the tangible – case studies, testimonials from satisfied customers and so on. This is a classic example of the need for being prepared for the objection in advance, by having suitable material that can be produced on the spot.

PERCEIVED LACK OF TRACK RECORD

As discussed in Section 4.8, although we may be convinced that we can apply our expertise in a new field, the client may not. As with the previous objection, we need to establish and demonstrate that we have worked successfully in the client's field, and have suitable material available to substantiate our claim. If we are making a first venture into the market, the winning of the first one or two contracts should be a specific aim and a high priority, probably requiring above average attention by marketing, scientists and management to increase the chance of success; immediate profitability is not the main objective.

LACK OF AUTHORITY

Our contact may seem to be happy but does not give the go-ahead. This may be because authority has to be sought from a more senior person or a committee.

What do we do if our contact says 'I'm happy with what you have said, but the decision will be made at the board meeting next Thursday week'? The answer is not 'ring on Friday week'! We should first ask whether we could take part in the presentation. Without this, we disappear from the scene and have to rely on our contact to do our selling for us, and it is unlikely that the presentation will be as convincing as the one we would give. Our contact is not an expert in our services; also, he or she may have been talking to competitors, and so their loyalty may not be as strong as we had believed.

If it is not acceptable for us to make the presentation, the next best plan would probably be for us to help our contact to do a good job on our behalf. We might provide additional handouts, overhead foils or other visual aids. We might help to work out the financial implications of the purchase, in terms of the return on investment for the company. After all, our contact wants to do a professional presentation to senior management, and should welcome anything we can do to help.

LOYALTY TO ANOTHER SUPPLIER

The potential buyer may seem to accept what we say but explains that they are using another service provider. We should not criticize the other supplier, because that would imply that the buyer had been exercising bad judgement. We might try to get part of the business at first. Multiple sourcing has many advantages, particularly in reducing risk and broadening the access to expertise. We might just drop the hint, in a low-key and uncritical way, that their present supplier may be taking them for granted. We could explain how we would provide all sorts of additional support and services that they are not currently receiving.

HIDDEN OBJECTIONS

There are times when everything seems to have gone well and we cannot understand why the buyer will not reach a decision. The problem may well be a 'hidden objection' which, somehow, we have to bring to the surface.

The open-ended questions described earlier can often reveal the hidden objection. If this fails, the next stage in the selling process – attempting to close the sale – may be the best approach (Exercise 8.2).

8.5 CLOSING THE SALE

There are literally dozens of techniques for closing the sale. They range from the low-key rational approach to those that border on trying to trick the customer into agreement. We concentrate in this book on those methods of closing a sale which can be used legitimately and honourably, and which are relevant to the selling of high technology services.

Our objective is that the buyer commits to going ahead with the deal (perhaps subject to contract). The most widely applicable techniques for closing a sale are described below.

THE TRIAL CLOSE

A trial close is not offering the prospective customer an opportunity to try out the service, as the title might suggest, because, unlike a hardware product, this is not normally possible with a service. The trial close is 'testing the water' in an unthreatening manner to see whether sufficient progress has been made.

A 'throw away' expression might be used such as 'are we ready to go ahead, then?' This approach may succeed. If not, the buyer may indicate areas that need more discussion, thereby revealing a hidden objection.

We should not give the impression that 'I must have an answer now'. In such circumstances, the buyer is likely to say that the answer is 'No!'

THE SUMMARY CLOSE

A summary close involves summarizing the progress made to date, orally, on paper or on a whiteboard. We note the various points we think have been agreed. At each one we carefully observe the reaction of the buyer, to detect positive or negative buying signals. The probability is that, faced with a written summary, the buyer will give the go-ahead. Alternatively, a hidden objection may be revealed which needs to be dealt with. The same or another method of closing can then be adopted.

THE ASSUMPTIVE CLOSE

The objective is to lead the buyer increasingly to think ahead as if the purchase had already been agreed. We would ask questions relating to the post-purchase period such as 'how frequently would you like meetings to monitor the project?' 'how many delegates will you be sending?', 'when will you want us to start?' and so on. Thinking about these issues may make the idea of having our service seem more tangible.

THE ALTERNATIVE CLOSE

This technique is one of the most common methods of closing and is used frequently in our personal and business life. In a shop it would take the form 'would you like the red one or the blue one?', 'will you take it or shall we deliver it?', 'cash or credit?' and so on. In an industrial environment there is considerably more scope for offering alternatives, not only in the service itself (the basic or the enhanced specification), but also in payment terms, timing, continuing support and so on.

The objective is to encourage the buyer to weigh up the relative merits of two alternatives, either of which is acceptable to us. The debate is 'yes versus yes' rather than 'yes versus no'.

THE CONCESSION CLOSE

A concession is offered in order to secure the sale. It might be better to offer free service rather than a straight financial discount, because it costs us less.

Also, the perceived value of what we are selling may be diminished if we are too ready to reduce the price.

There are two important points about the concession close. First, we should not go as far in the first instance as we might be prepared to do. If we are prepared to offer 10 per cent discount, we might offer 5 per cent and allow ourselves to be talked up to 6 or 7 per cent; the buyer has won a victory, and we have saved some money. Second, and even more importantly, we should trade the concession for something, preferably the contract. We might use an expression such as 'if I were to give you a 5 per cent discount, or include that extra service without charge, would you confirm the order now?'

KEEPING CONTROL

Whatever method is used for closing the sale, the seller must keep control of the process. We cannot wait for the sale to close itself. How often do we fail to make a purchase when an intelligent and sensitive use of an appropriate closing technique might have been successful?

By the way, the sale has not been completed until we have received the money! (Exercise 8.3.)

8.6   THE SELLING ROLE OF NON-SALES STAFF

Many companies have a clearly defined department responsible for the selling function. However, it is quite common for organizations selling high technology services not to have a single member of staff whose main role is selling. The only people available for selling are those who are employed primarily to carry out projects and generate fees.

Some operations fall somewhere between these two descriptions – they have a small sales resource which they know is inadequate, but they are not sure whether to expand this or to ask non-sales staff to do some selling on a part-time basis.

By 'sales staff' we do not mean unqualified sales representatives or redundant staff from a different field. They will almost certainly have a degree or professional qualification and possibly a PhD in a relevant scientific or engineering discipline.

There are advantages and disadvantages to both approaches. Some of the factors are listed below.

NON-SALES STAFF: ADVANTAGES

1.   They can make the service more tangible. They live in the world of the technology and are able to speak about it with authority.

2.   The needs can be thoroughly investigated. They realize the finer points of detail about the service and its application, and should be in a position to ask careful probing questions.

3.   They can establish a greater rapport with the client. This is particularly true when they are speaking to other high technologists who understand the jargon and ways of thinking which are common to both of them.

4.   Quality is more easily demonstrated – the seller is the 'product'. 'What you see is what you get'.

NON-SALES STAFF: DISADVANTAGES

1.   The selling effort is limited to the availability of people who are actually employed to do something else.

2.   Non-sales staff can waste a great deal of time in their enthusiasm for the technical aspects of the product.

3.   Most non-sales staff are technically oriented rather than customer oriented, and live in a world of features rather than benefits.

4.   With intangible services such as consultancy or research, the 'seller' may be so enthusiastic as to reveal key parts of the solution to the buyer during the selling process. The buyer goes away highly delighted, solves the problem and is never seen again.

FULL-TIME SALES STAFF: ADVANTAGES

1.   More sales effort is available. Staff are employed full time for this task, and are motivated and usually given incentives to do nothing but win contracts.

2.   They have better control over customers. They live in their world and understand how they make decisions.

3.   They should have a market orientation rather than a product orientation.

FULL-TIME SALES STAFF: DISADVANTAGES

1.   There may be a high cost per customer because they quickly find that they need to call in a specialist to back up their own efforts.

2.   Sales people *may* have an unnecessary tendency to price cutting. If their commission depends on turnover rather than profit and they are authorized to give discounts, why would they not do so? This is a failure of management, but it certainly occurs.

3.   Sales staff may put forward unnecessarily customized solutions. They are so anxious to get the sale that they concede points on the specification that have more far-reaching technical and cost implications than they had realized.

There are clear advantages and disadvantages of using full-time and part-time staff in the selling operation. The answer for an organization selling high technology services may lie in a combination of both (as is the case with many of my own clients), along the following lines:

1.   Use impersonal communications early in the sales cycle to identify qualified leads.

2.   Use full time generalist sales staff to follow up those leads to achieve further qualification.

3.   Bring in specialist high technology experts in those cases where a sale is most likely. In this way we can enjoy the advantages of both options without the disadvantages (Exercise 8.4). It is a clear example of 'selling the next step'.

8.7 NURTURING EXISTING CUSTOMERS

Nurturing existing customers may not seem to bring incremental short-term rewards, but it is an essential part of our task. Once we start trying to win new customers, we realize how much less it costs to deal with our existing ones, so we need to protect our position as their historic supplier. We cannot take anyone's loyalty for granted, and competitors are attempting to win our customers for themselves. In any case, there are other reasons for looking after them. They may speak spontaneously on our behalf, or they may be willing, for example, to open their facilities as a 'reference site' to which we can invite potential customers. Such privileges should not be abused or used too often, but they can be an invaluable method of reinforcing our own marketing efforts. After all, we are paid to say that we are good; customers are not, and their testimony is therefore much more credible.

An interesting issue arises with a dissatisfied customer. It has been said that a satisfied customer can bring us five more, and a dissatisfied customer can lose us ten. We should deal with the cause of dissatisfaction if at all possible, because it can influence other potential buyers. This is just as true if the fault is not actually ours; hearers of the dissatisfaction will react as if it were. However, and this is the main point, a dissatisfied customer whose cause of dissatisfaction has been removed may actually become more loyal than someone who has never had cause to complain. A vaguely satisfied customer may, in fact, not be particularly loyal; a customer who has had a problem that the supplier went to great lengths to solve may become one of that supplier's most positive allies in the future.

We need to strike a careful balance in our selling efforts between existing and new customers (Exercise 8.5).

8.8 PARTNERSHIPS AND STRATEGIC ALLIANCES

Some companies have a world-wide network of subsidiaries through whom they do business. Companies selling hardware products may need a network of operations that holds stock and spares, employs full-time sales and service staff, and has the financial and administrative infrastructure necessary for doing business in their region.

For most organisations selling services, particularly small ones, this sort of structure could not be justified. The good news is that it is not necessary.

Although trivial compared with the business of many readers, the way in which I have carried out seminar business in some Far Eastern countries, from the UK, is a good illustration of what can be done with minimal commitment. The secret lies in having the right partner. In my case it is a member of a university who runs seminars for high technologists and has a large database of potential delegates across the Pacific area. However, he did not have a seminar on marketing suitable for high technologists. He had market access, but no 'product'. I had the product but no market access. The combination of the two of us, in a contractual but relatively informal partnership, was all that was needed. Both partners have invested time and money in the venture, but at a very low level compared with what would have been necessary if I were trying to do the whole job without a partner.

Much larger companies are increasingly moving towards strategic alliances with other large companies for their international business; the principle is the same.

However, a number of such partnerships do not work, and we need to ask why. From my experience there are some important principles which need to be followed for success to be achieved. They lie first in selecting the right partner and then in managing the partnership to achieve the commonly agreed objectives. What are these principles?

1.   The partner, whether an individual or a company, should have the right profile. This involves qualifications and experience, and a clear means of market access.

2.   The partner should have good motivation. This can make all the difference between success and failure. The partner should be 'hungry' for business on our behalf. This means that they will be properly rewarded for success, and will need us at least as much as we need them. They may be working for other (non-competing) companies, but have committed to giving us a sufficient level of their time and effort. Some people who complain that their overseas partners are not doing a good job have given them business terms that virtually guarantee failure; unless a partner can see a good prospect of sharing in the financial success, why should they bother?

3.   We will support them fully and expect results. This may mean training them, sitting down with them to agree business plans for our service in their region and being readily available to follow up any leads they obtain on our behalf.

4.   We will concentrate our efforts on key markets, rather than dissipating them thinly across the world. We will have plans to exploit these markets, and will ensure that sufficient resources are committed to them. If we are successful, we can use these successes as stepping-stones to further internationalization, but we will control the speed at which we attempt to do this.

5.   We must understand that different parts of the world have significantly different national, business and financial cultures. One of the benefits of having a partner who is a national of his or her own territory is that they understand these matters in a way which it is very difficult for a 'foreigner' to do.

If we adhere to these principles, as long as there is genuine potential for our services in their region, we should be able to obtain good business with relatively little expense and risk compared with many other ways of doing business.

Although these partnerships and strategic alliances have been discussed in terms of international business, there may be situations where a partner in our own country can also be appropriate; the principles for successful partnership are the same.

The same applies when we want to enter a new market segment or a new field of technology. The right partner can often enable us to achieve more and to do it more quickly and more cheaply (see Exercise 8.6).

EXERCISES

8.1 Think of a sales opportunity for your company. What do you want to know about the client? What open-ended questions could you ask in order to elicit this information?

8.2 Make a list of the main objections you encounter in selling your company's services. Against each, write the main points in your argument for overcoming it.

8.3 How would you use the assumptive close, the alternative close and the concession close in your business?

8.4 Consider the advantages and disadvantages of using non-sales staff and full-time sales staff in your business. Do you think you have the right balance?

8.5 Are you paying enough attention to existing customers? Consider some customers who have recently been dissatisfied with your company. Do you think the dissatisfaction was handled in the best way? How could it have been improved? Can you do something about it now?

8.6 Should you consider setting up one or more partnerships overseas? If you already have some, are they working well? Have you adopted the five principles suggested? Could the same approach be applied to a new market segment or a new area of technology which you are considering entering?

FOLLOW-UP

8.1 Make a habit of writing down two or three questions in advance of each meeting with a potential customer.

8.2 Make a habit of considering items 1 to 5 of Section 8.3 in advance of each presentation.

9

HOW DO WE USE IMPERSONAL COMMUNICATIONS?
LITERATURE, PROPOSALS, WEB SITES, LETTERS, MAILSHOTS, E-MAIL, TELEPHONE, EXHIBITIONS, ADVERTISING AND PUBLIC RELATIONS

Key business issues	*Section*
◯ Much promotional literature is ineffective because it relates more to the supplier than to the customer	9.1
◯ Promotional literature can often be significantly improved in a relatively short time if it is judged against certain guidelines by an in-house panel	
◯ Literature often attempts to reach too many different targets	
◯ Proposals and responses to invitations to tender often fail to differentiate the supplier from other competitors. It is essential to identify, emphasise and develop the 'win themes'	9.2

○ A proposal can often be significantly improved in a
relatively short time if it is judged against certain
guidelines by an in-house panel

○ Most web sites only contain information about the
supplier and the services offered, and do little to
promote the services or identify with the needs of the
customer. Most of them would benefit from being
rewritten   9.3

○ Sales letters, mailshots and e-mail are often not read
because insufficient thought has been given to the
message and the response expected   9.4

○ A well-designed reply form can increase their
effectiveness

○ Use of the telephone for marketing can be made much
more effective by following simple principles   9.5

○ Much of the money spent on exhibitions is wasted.
Properly planned and used, exhibitions can be one of
the most cost-effective tools of marketing communication   9.6

○ Qualification of leads is the key to exhibition follow-up

○ Much of the money spent on advertising is wasted
because its purpose is not clearly understood   9.7

○ Public relations (PR) creates a background against
which other marketing efforts can become more
effective. Public relations is often under-utilized by high
technology service organizations   9.8

9.1 PROMOTIONAL LITERATURE

THE CIRCUMSTANCES IN WHICH LITERATURE IS RECEIVED

Imagine the scene. Someone is expecting to receive a piece of our literature
through the post that morning. He wakes up and says to his wife, 'I'm looking
forward to today – XYZ are sending me their new brochure'! He arrives at the
office and enquires eagerly whether the post has arrived. He asks his secretary
to cancel his appointments and take his telephone calls so that he will not be

disturbed, clears his desk and spends two hours reading the brochure from cover to cover.

If only that were so! In reality, the document first has to get past the secretary, who is trained to protect the manager from junk mail and other time-wasting intrusions. It then arrives in the in-tray, where it quickly becomes covered with other documents. It sits there for several days until the recipient has a blitz on the pile, after which it will end up in the bin, in the filing tray or – and this is probably the best we can hope for – in the briefcase to be read at home or on the train. In other words, we have about two seconds to capture the recipient's interest, and to create a greater impact than all the other documents which constantly flood in.

One of my favourite photographs is of my young son lying on a beach in France devouring a model railway catalogue – which he did for a whole fortnight. If only our literature could command the same degree of attention!

How are we to do this? I make no apology for repeating the first three simple questions posed in Section 7.1:

1.   *What are we trying to say?* Is our message concerned with establishing an image, giving information, creating warm feelings, generating a demand or is there some other message?

2.   *To whom?* Will it be read by specialists or by generalists or both? Will it be read by people whose interest is primarily technical, financial, operational or what?

3.   *With what objective?* Is the document merely intended to supplement other marketing efforts, or is it aimed at getting a specific response in its own right such as a request for further information, a telephone discussion, a visit, a demonstration, an invitation to bid or even an order?

Some guidelines for effective promotional literature are given below. It is a useful exercise to take a piece of our existing literature and, very objectively, award marks against these criteria. I do this exercise on most of my in-company seminars, and it is surprising how often the marks turn out to be low, even for a piece of literature with which the client was previously quite happy.

PROMOTIONAL LITERATURE GUIDELINES

Suggested guidelines are:

1.   *Impact on first sight.* What is the immediate impression on seeing the front cover of the document (within a maximum of two seconds)? Is

it going to create sufficient impact to ensure that it will at least be kept for later reading?

2.   *Initial interest.* Does it capture the recipient's initial interest compared with all the other documents calling for attention, so that it will be read further when there is time? In the meantime, will it leave some sort of positive impression?

3.   *Customer orientation.* Does the document relate to the needs of the reader, does it present benefits and are these benefits targeted to the needs of the particular recipient, or does the document just talk about the supplier and its services?

4.   *Uniqueness.* Does it establish a unique selling proposition? Does it present the company as offering a unique service, or at least a special service, to differentiate it from the large number of other companies offering apparently similar services?

5.   *Logical structure.* Does it take readers logically through a progressive argument leading to a conclusion and action, and can they easily find the parts they might want to study?

6.   *Continuing interest.* Is it sufficiently interesting to sustain attention, so that the reader will be willing to devote some time to studying it in detail?

7.   *Market segmentation.* Which market segments is it aimed at? Does it do this adequately, or does it 'fall between several stools'?

8.   *Decision-making group.* Is it directed at members from several different disciplines (technical, financial, general management, operations, quality etc.)? If so, is there at least something for all of them (even if some parts are more relevant than others). Can they find the part that they might want to read, or is it hidden in a mass of detail which, to them, is irrelevant?

9.   *Comparison with competitive literature.* Will our proposed document compete favourably when laid alongside current copies of the competitive literature? If a potential client were compiling a short list of consultants from whom to request a proposal, would we pass this first stage? We often have to win the 'battle of the literature' before we can win the 'battle of the product'.

10.   *Novelty.* Does the document have something to lift it out of the ordinary, perhaps by really creative use of photography, diagrams, pictures of computer screens and so on. Most companies have buildings with windows and cars parked outside – we have to do better than that. A promotional brochure is not a catalogue – it is a document fighting for attention.

11.   *Style and image.* Is the style appropriate to the marketplace, the reader's operation, the supplier's situation and the service we are trying to promote? Quite apart from the actual content, what does the reader perceive about the supplying company from the quality of the document and the image it conveys? One client admitted that their literature very accurately portrayed their company's image – conservative, short of money, out of date and boring!

12.   *End of the document.* Does it reach a conclusion and invite a response, or does it simply stop because there is no more space? Does it contain some sort of offer, possibly for further information or for a visit or demonstration? For example, the brochure promoting my own seminars invites readers to telephone for an informal discussion; in this way, they are able to consider the relevance of the seminars without any commitment or high pressure selling. Should the name of the contact be printed? If it is, the person usually leaves the next week! If it is not, it is impersonal. A slot for a visiting card is a useful way of resolving this dilemma.

Examining a document against these 12 criteria will often reveal some specific steps which can be taken to improve its effectiveness next time it is printed. While it is not suggested that documents can be written by a committee, frequent use of these criteria on workshops has shown that a small group of non-marketing managers who have spent a few hours learning the basics of marketing can produce very sensible proposals for improving the effectiveness of a piece of literature in a surprisingly short time. A 'marking sheet' based on the above criteria takes the following form (Exercise 9.1):

Criterion	*Marks out of 5*
Impact on first sight	_____
Initial interest	_____
Customer orientation (benefits)	_____
Uniqueness	_____

Logical structure   _____

Continuing interest   _____

Market segmentation   _____

Decision-making group   _____

Comparison with competitive literature   _____

Novelty   _____

Style and image   _____

End of the document   _____

Total (out of 60)   _____

THE PROBLEM OF MULTI-PURPOSE LITERATURE

Many documents fail to be effective because they attempt to be 'multi-purpose'. They are directed at too many targets, and therefore do not address any targets adequately. This often comes about because the whole budget is spent on one rather lengthy document, whereas it might be far better to produce a number of one-page leaflets each of which has a specific and quite different purpose.

Even then, the instinct is for companies to write product-oriented literature and to use it for all market segments. It might be better to have market-oriented literature, each piece covering a number of services relevant to that particular market.

It is worth planning a hierarchy of literature, descending from the general to the particular rather like a family tree. It is not necessary to produce all the components at once, but each new addition will fit logically into a whole scheme.

A cost-effective strategy might be to have a good glossy general piece of literature designed to establish credibility, reputation, image and so on, and then to have a number of different short leaflets each targeted at a particular market segment. It might be acceptable for these short leaflets to be produced internally by desktop publishing; this has the great advantage that they can be updated frequently as the track record develops and even modified to meet particular circumstances. With a little ingenuity, the general brochure can be printed in the form of a wallet that can contain an appropriate selection of the individual market segment leaflets. In this way the literature budget can be used to best effect, and individual customers can be given the documents which will be most effective in identifying with their particular needs (Exercise 9.2).

9.2   PROPOSALS AND INVITATIONS TO TENDER

Many companies are accustomed to a culture in which they produce 'estimates' or 'quotations' in response to 'enquiries'. These documents do not attempt to make claims about what they are selling, to differentiate themselves from the competition, or to persuade by producing business arguments. They are typical of a reactive marketing approach. These companies have to learn instead to produce 'business proposals' in response to 'marketplace opportunities'.

Two different examples may be distinguished. In the first case, we are completely free to decide the format, length, content and emphasis of the proposal we are preparing for a potential customer. In the second case we are given a proscribed format, sometimes extremely lengthy and complex in its structure, which we are required to complete; failure to comply with the required format may automatically invalidate our tender submission.

In either of these cases, it is strongly recommended that, before beginning to draft the sections and 'fill in the boxes', we write down on a single sheet of paper the key points which we judge will constitute the 'win themes'.

This is an important discipline for two reasons. First, if we cannot briefly encapsulate the message we are wishing to convey, it is doubtful whether we have sufficiently thought it through. Second, the more the client is wanting to push our submission into their format, the more effort we will have to make to ensure that we emphasize the points that *we* want to communicate (while still complying with their requests). There is the story of the preacher who said 'First I tell them what I am going to say, then I say it, then I tell them what I have said'. This is good advice for the submission of proposals or, indeed, for most forms of marketing communication, because it clarifies and reinforces the message to be communicated.

Before we begin to write a single word, we need to know who will read the document. Will the award be adjudicated by an individual or by a board or committee? In the latter case, which business disciplines will the group represent? (See Decision-making group in Section 2.6.)

We should be very clear as to the specific objective of the proposal. Is it intended to achieve the award of the contract in its own right, or is there a preliminary objective such as a demonstration, a feasibility study, a visit, prequalification or qualification for the short list? We should put all our effort into achieving the specific objective of the document and not do a half-hearted job of trying to win the contract if this is not the likely outcome of this stage in the negotiating process.

We then need to identify the competitive situation within which this particular award will be considered (see Section 5.1). If we have not correctly understood this, we may 'win the battle and lose the war', i.e. convince the reader that our service is the best, but lose the contract to some other form of competition (such as an alternative way of solving the problem, or the client doing the work with internal resources).

We are now in a position to start writing the document. The one-page summary of the 'win themes' should be prominently displayed (perhaps on a whiteboard or flipchart) while we or our colleagues are writing the text.

Some suggested criteria against which the proposal can be assessed are shown below. Some of them are similar to the criteria for assessing promotional literature, because many of the needs are the same.

1. *Summary.* Are the key issues and 'win-themes' clearly set out in an introduction or executive summary? Many of the recipients will be too busy to read the whole submission, and will form their judgement on the basis of the summary and a selected inspection of parts of the main proposal. Our aim should be to obtain virtual agreement from the summary alone; we do not necessarily *want* everyone to read all the detail. The summary sometimes takes the form of a covering letter, but this should be avoided if there is any risk of it becoming separated from the main proposal.

2. *Impact on first sight.* Most proposals and tender submissions are likely to receive at least a cursory glance. However, we have to bear in mind the psychology of the process and the fact that those who award tenders are still human beings with all the normal motivations and prejudices. If a proposal looks professional and well presented, it is more likely to be approached in a favourable light than one that looks boring or confusing.

3. *Continuing interest.* Does it look interesting and worth reading at length? Does it contain pictures, diagrams and arresting statements at strategic points in the document so that, if the reader glances ahead, the appetite will be whetted?

4. *Competitive situation.* What form of competition is the proposal designed to address? Does it do this effectively?

5. *Customer orientation.* Does the document relate to the reader's needs by highlighting relevant benefits, or does it just talk about the supplier's services, skills and resources?

6. *Decision-making group.* If the document is addressing members from several different disciplines (technical, financial, general management, operations etc.), is there at least one part that is designed to attract each and every member of the decision-making

group? We can lose a tender by failing to convince only one member of the DMG.

7. *Uniqueness.* Does the proposal present the company as offering something unique, or at least special, to distinguish it from the competition? As we explained in Chapter 2, the differentiation may be in the form of intangible, peripheral or personal attributes and not necessarily in the service itself.

8. *Curricula vitae* (CV) – if appropriate. Have CVs been specifically targeted to this project and the needs of this particular client? Obviously we cannot alter the basic content of a CV, but the emphasis and method of presentation can make a great deal of difference. It is much easier to pull a standard CV off a word processor, but it is worth spending a few moments targeting it. This is particularly true of consultancy, where the skills and experience of the consultants are paramount. It might be possible, for example, to highlight a panel showing 'experience of Dr X which is particularly relevant to this project'.

9. *Financial issues.* Has a soundly based financial argument been presented? This may involve issues such as investment and return, lifetime cost, risk and sensitivity (Section 11.5) and cash flow, in order to assist the senior manager or finance-oriented reader to reach a decision. A detailed cost breakdown, which is the only financial information given in many proposals, does not begin to address this subject.

10. *Structure and layout.* Has full use been made of the freedom which the client permits to present the argument in the most persuasive manner, or have the 'mechanics' got in the way? Does the document take the reader logically through a progressive argument leading to conclusion and action? Does it have a good index (if appropriate)? Can the sections be found easily? Again we are recognizing the fact that people are busy (or even lazy), and we want to make it as easy as possible for them to receive our message.

11. *Style and image which it conveys.* Is the style appropriate both to the buyer and to the seller? Has good use been made of graphics, diagrams, charts, graphs etc., in a way which conveys the appropriate degree of professionalism? Quite apart from the actual content, what does the reader perceive about the company from the quality and style of the document? The unwritten message (professionalism, or

'this bidder really understands our requirements') can sometimes be more powerful than the written message, and images which the recipient receives from 'reading between the lines' can speak more powerfully than the words themselves.

12.   *Key issues.* Are the 'win themes' clearly developed throughout the text, and are they emphasised and progressively supported? If practicable, is there a recapitulation at the end that leaves the main points in the reader's mind? Remember the preacher!

A 'marking sheet' based on the above criteria takes the following form (Exercise 9.3), (Follow-up 9.1):

Criterion	*Marks out of 5*
Summary	_____
Impact on first sight	_____
Continuing interest	_____
Competitive situation	_____
Customer orientation (benefits)	_____
Decision-making group	_____
Uniqueness	_____
Curricula vitae	_____
Financial issues	_____
Structure and layout	_____
Style and image	_____
Key issues	_____
Total (out of 60)	_____

9.3 WEB SITES

One of the most common actions which delegates agree to take after one of my seminars is to restructure and rewrite their web sites. Many web sites simply present the same material as the company's brochures and should therefore be subject to the same scrutiny as described in Section 9.1. Many of them exemplify the description 'this is who we are and this is what we do', which was criticized in Chapter 1 and elsewhere.

However there is a further aspect of web sites which makes it even more necessary to think very carefully about the structure; they have an almost infinite capacity for complexity, which means that they have an almost infinite capacity for causing confusion.

Two recent exercises with groups of delegates illustrated the problem quite dramatically. In the first, the main conclusion they came to was that they could not even find their way round their own web site. Customers are not likely to persevere if this is the case.

In the second, the conclusion was even more bizarre. They were being asked 'could the serious seeker or the random browser easily find what they wanted?' The conclusion was that the random browser might, but the serious seeker would not. The reason lay in the fact that the optional pages which could be accessed were defined by key words which the writers understood but which were not the obvious ones to come to the mind of the readers. Search engines have amazing capabilities but, if the key words are not selected with the customers' needs and vocabulary in mind, they are likely to find either tens of thousands of references or none at all.

The large capacity of a web site offers the ability to produce targeted material to a much greater degree and at a negligible cost compared with that of the same amount of printed literature. This medium can also be updated rapidly and frequently if necessary, at virtually no cost apart from the time taken.

In Section 9.1 we thought about a hierarchy of literature, descending from the general to the particular rather like a family tree. Surely that is exactly what happens in a web site – we log on to the home page, and then follow various paths indicated by icons, key words or buttons to reach the subsidiary pages. This parallelism between literature and web sites suggests that it might be worth considering both as parts of the same exercise rather than as separate exercises. They need not contain the same text, but they may well have the same logical structure; if, therefore, we conceive both at the same time, we might achieve consistency in our communications and save a great deal of time in the process (Exercise 9.4).

9.4 SALES LETTERS, MAILSHOTS AND E-MAIL

In this section we are referring to any written communication with a potential or existing customer that is designed to further the communications process and qualify the lead as described in Section 7.1. These might include letters to individuals with whom we have had prior contact, or a programme of 'cold mailshots' to people whose names and addresses we have obtained in some way.

A number of the points below have already been made in connection with other forms of communication. This is quite deliberate, and the reader is encouraged to take a common approach to the whole subject of marketing communication.

The following guidelines might be used for judging a letter that has already been written, in order to clarify our thinking for the future. It might be one that we have sent to a client, or one that we have received from someone attempting to sell to us.

Two points are paramount:

1.   We must be clear about the objective of writing, and emphasize this in the opening and closing paragraphs. The entire letter must be worded in a way which is most likely to achieve that objective. For example, if the specific objective is to 'sell' a meeting, the emphasis should be on the reasons for having a meeting; generalized messages about the company or the service should only be used as a means of reinforcing the main aim.

2.   We need to find some way of getting past the secretary who is paid to protect the manager from 'junk mail'. The letter should always be addressed to the reader by the correct name (which has been checked by telephone if necessary) – never 'The Managing Director: Dear Sir or Madam'. The opening sentence should refer to a specific event such as a meeting or project, or describe financial or technical issues in such a way that the secretary does not feel authorized to reject it.

GUIDELINES FOR A SALES LETTER

1.   *Objective*
       a)   Why are we writing? (General objective)
       b)   What should the letter achieve? (Specific objective, e.g. next step)
2.   *Structure*
       a)   Opening:
              i)   Commands attention.
              ii)   States the objective.
              iii) Establishes link with previous contact or event.
       b)   Middle:
              i)   Develops the theme(s).
              ii)   Presents facts.
              iii) Offers benefits.
              iv) Creates a link to the close.

    c) End:
        i) Summarizes.
        ii) Clearly states the next actions.
3.   *Layout*
    a) The letter should have good visual appeal.
    b) The eye should travel smoothly down the page.
    c) It should be easy to see which parts refer to what.
4.   *Language*
    a) Avoid trite and over-used phrases.
    b) Avoid pomposity.
    c) Avoid both coldness and inappropriate familiarity.
    d) Ensure perfect punctuation, spelling, use of tenses.
5.   *General impression.* The whole letter must be:
    a) attractive
    b) readable
    c) credible
    d) persuasive, and must
    e) represent our company as a worthy partner (Exercise 9.5).

E-mail will normally be shorter than letters, but the same logic should apply.

REPLY FORM, USED MAINLY FOR MAILSHOTS

One of our objectives is to make it as easy as possible for the recipient to respond. It may be worth including a reply form. The example in Figure 9.1 was used in the case study described in Section 7.2; the reader might like to reread that section in order to understand the circumstances in which the reply form was used.

Some important points need to be made in relation to this reply form.

The first option in Figure 9.1, 'Please contact my secretary to arrange to visit me ...', is based on an absolutely crucial point. Why do we want to arrange the meeting through the secretary and not the manager himself or herself? The obvious answer – that the manager is busy – completely misses the point. What is the objective of the letter? It is to gain an interview. At what point have we achieved our objective? When the manager *has ticked the first box*. Once we have achieved this objective, we must not allow a change of mind. We do not want to start discussions over the telephone, because our aim is to meet the buyer face to face. Once the first box has been ticked, the secretary will not feel able to question the decision. The manager has asked for an appointment, and all that has to be done is to arrange a mutually convenient date.

If I were a purist, there would only be the one box to tick, and there would

To: (*ourself*)

(please tick as appropriate)

[   ] Please contact my secretary to arrange to visit me for an exploratory
        discussion

[   ] Please telephone me to discuss the possibilities in more detail

[   ] Please send me more general information

[   ] Please send me specific information in the areas shown below

[   ] Please contact me again in .... months' time

[   ] Please delete me from your mailing list

Other comments, or areas of interest :

Name .... *(filled in beforehand by ourselves if possible)*

Company ....

Position ....

Figure 9.1   Reply form

be no other options. However, I am also a realist, and so some alternative options have been offered.

The last option, 'Please delete me from your mailing list' is controversial and should only be used where the target market is effectively infinite. It is based on the assumption that an initial rejection is a useful piece of information and that our efforts would be better directed to other potential clients on the list. It is a negatively qualified lead.

The response can be increased if we enclose an envelope addressed to ourself with a stamp on it – not a business reply envelope. This is based on the assumption that people will not throw stamps into the bin in the way that they would a reply paid envelope; the easiest way of dealing with it is to tick a box and put it in the out-tray.

It is not suggested that this particular approach is a stereotype for all business situations, but I have used it very successfully myself and it may give food for thought to other readers (Follow-up 9.2).

9.5 THE TELEPHONE

The telephone is increasingly becoming a vehicle for marketing communication. Its use is extended by fax and video conferencing, and no doubt we shall soon be communicating by virtual reality!

The telephone involves much lower cost and management time than most other forms of communication. If used properly, as one step in the communications process (see Section 7.1), it can be very effective.

However, the telephone can also be extremely frustrating and even counter-productive if certain tactics are not followed. It can easily be ignored, or a message can be rejected without any embarrassment on the part of the recipient. It is much harder to gauge the reaction of a customer over the telephone than in a face-to-face meetimg where body language and behaviour can be very significant.

We can distinguish two situations. When the caller is known, the conversation is relatively straightforward. When the caller is unknown, there is an added complication and a higher risk that the call will go wrong. We will therefore consider the more difficult of the two – the 'cold call'.

As with all forms of marketing communication we need to decide in advance what is the objective of the call. For most readers it is quite unrealistic to expect to gain an order at the first contact over the telephone; the purpose of the call is to try to diagnose the potential for doing business and, if this is favourable, to move on to the next stage of the communications process – arranging a visit, for example.

The first step is to get past the switchboard without discussing the reason for the call. In a large company this is not normally a problem, but if we are trying to contact someone in a small company or in the home, the person who answers may say 'X is not here at the moment – can I help you, or would you like to leave a message?' That is the last thing we should do. The call may not be returned, but that is not the point. If it is, it can place us at a serious disadvantage. We should spend at least a few moments preparing for the call, recalling its objective and refreshing our memory on the background by referring to previous documents if necessary. If we receive a return call 'out of the blue', we may be in a meeting, we may forget why we wanted to speak to this person or, at very least, perform less well than if we had made the call ourselves after suitable preparation. In other words, we want to take and keep the initiative.

Let us suppose that we have been put through by the switchboard to the target's secretary. Part of the secretary's job is to save the manager's time by filtering out unwanted callers. We must somehow make the secretary feel unable to reject the call or to make any decision about the subject matter. We might refer to important financial issues or to specialist or technical details that the secretary does not feel able to comment upon.

What do we do when we have overcome the first two hurdles and are now speaking to the right person? Let us suppose that we are trying to arrange a visit. The purpose of that visit will be to diagnose the potential, start to unearth the customer's problems and begin to propose a solution. Our aim is that these key steps should be left to the meeting and not dealt with inadequately over the phone. If we are going to fail, as we sometimes will, we would rather fail after a face-to-face meeting. Marketing a meeting is no different from marketing a service – we have to sell the benefits of the meeting.

This means that we will deliberately try to resist requests to 'tell me a bit more about what you are selling'. We are trying to sell the interview not the service. We therefore have to try to persuade the hearer to give us ten minutes or two hours or whatever is appropriate, at a meeting in their office.

They may try to reject the idea of a meeting by asking 'could you send me some literature?' This is probably not a genuine request, but a polite way of saying 'no'; it should therefore be resisted, possibly along the lines 'I could send you some general literature, but if I can just visit you briefly I can give you the literature which will be most helpful to you, perhaps some case studies relating to your situation'.

Properly used, and bearing in mind the multi-stage approach to communication, the telephone can be a tremendously useful and cost-effective means of marketing communication. Badly used, it can be an expensive way of losing business (Follow-up 9.3).

9.6   EXHIBITIONS

THE ROLE OF EXHIBITIONS IN THE SELLING PROCESS

Experience of running exhibitions of all sizes in all parts of the world has led to one clear conclusion. Exhibitions can be the best and the worst way of spending marketing money.

Exhibitions can cost tens or hundreds of thousands of pounds. On the other hand, as a management consultant, I have taken a few square metres at a large international exhibition for less than £2000, and at a more specialized exhibition in a hotel for a few hundred pounds. Those who have dismissed exhibitions as being beyond their reach should perhaps think again.

An exhibition can take a potential buyer a long way along the selling process – sometimes from beginning to end – all on the one occasion.

The stand and its display panels, the literature, and our mere presence at the exhibition are forms of impersonal communication aimed at creating awareness and generating interest. A visitor who has shown some interest in the 'impersonal' material can then be approached with 'personal' communication (see Section 7.3).

An exhibition is a classic case of 'selling the next step' (see Section 7.2). This step may be a written proposal, a visit to the client, a visit to the seller and so on. We need to qualify leads by follow-up category – we do not want quantity of leads but quality. An unqualified list of names and addresses is useless because there is no option but to treat them all in the same way – putting them on a standard mailing list.

TRAINING FOR STAFFING A STAND

Everyone staffing the stand should be trained in the basic principles of marketing, most simply encapsulated in the diagram on the 'sharper cutting edge' (see Figure 2.2 in Section 2.8). The instinct of many people, faced with a visitor to the stand, is immediately to talk about the services on display and to explain what the company does. Staff must be trained to present benefits rather than features, to target those benefits to the needs of each individual visitor and to present the company's unique selling proposition. These presentations can only be done if some diagnostic questions have first been asked, so that they understand at least something about the possible needs of each visitor.

Similarly, the exhibition panels should concentrate on the likely needs of the buyer, and not only on the services of the seller as they usually do.

ADVANTAGES OF EXHIBITIONS

Those who are unsure about the cost-effectiveness of exhibitions should compare them with alternative ways of spending the same amount of marketing budget – more literature, more visits by sales people, more advertising and so on. If we do this, we shall probably conclude that there are some distinct advantages afforded by exhibitions.

1. *Potential customers come to the seller.* This saves us time and money, particularly at an international exhibition.

2. *Many customers can be seen in a short time.* We can hold several

useful business conversations per hour, compared with one or two per day if we were to visit buyers at their own locations (or fewer still overseas).

3.   *Responsible decision-makers attend and are accessible.* This includes people who are not normally available to be seen during the selling process, such as senior managers. At an exhibition they are available, relaxed, more willing to talk and even to open their diaries.

4.   *New potential customers attend.* People who would never have been on our long list let alone our short list of potential customers may be contacted at no incremental cost.

5.   *Contact is face to face.* More can be achieved in an hour at an exhibition than in many days of patient plodding by other means.

6.   *Exhibitions offer good visual impact.* People who would never respond to an advertisement or a piece of literature may be attracted by a stand and its display.

7.   *An exhibition can be a good vehicle for a 'mini test market' of a new service.* Nowhere else can we encounter such a large group of potential buyers in such a short time. We must beware of the problem of an unrepresentative sample (see Chapter 10), but that is true of any market research.

8.   *An exhibition is an excellent vehicle for evaluating a new market segment.* Many companies are having to diversify into new areas, and an exhibition can act as a very good first filter. We might decide to go no further because competition is intense, or we might judge that the potential is sufficiently great to justify further investigation; alternative ways of reaching these conclusions might take much longer.

9.   *A large number of leads can be generated and qualified for follow-up.* The work starts when the exhibition is over.

DISADVANTAGES OF EXHIBITIONS

Compared with other forms of marketing communication, exhibitions have some potentially serious disadvantages:

1.   *They are expensive.* The stand space, in the form of a 'shell scheme' to which one has to add panels and possibly a whole internal

display structure, can cost more than £200 per square metre. On top of this there is the cost of fitting and providing services. The cost of staff is more than their salaries, travel, accommodation and related expenses – it is their 'opportunity cost'. While they are attending the exhibition, they cannot be earning fees or generating profit in other ways.

2. *Much of the effort is wasted.* It is very difficult to avoid spending time with people who do not buy. However, a large proportion of any marketing effort is wasted, and an exhibition can actually be very productive.

3. *The environment is very competitive.* A buyer whom we visit may or may not see some of our competitors before making a decision. A buyer visiting our stand can evaluate the whole of the competition the same day.

4. *Instant judgement is required.* Unfortunately, the people who express most interest in the services – students or R&D staff who are trying to keep up with the latest technology – may be the ones with the least ability to place a contract. We need to decide how best to allocate our precious time.

5. *Exhibitions can be extremely useful to competitors.* They will visit our stand (having conveniently mislaid or changed their badges!), in order to find out the latest technical and commercial information.

OBJECTIVES OF TAKING AN EXHIBITION STAND

Objectives will vary widely, and must be well thought out, agreed and understood by all concerned before the exhibition takes place. In many exhibitions, in the USA for example, visitors come prepared to buy. They expect special exhibition offers. In other countries such as Germany, the whole environment may be much more restrained and little business may be concluded at the exhibition itself.

Some possible objectives of taking a stand are:

1. *To obtain orders.* If it is the sort of exhibition where this is possible, the right staff and information must be available.

2. *To obtain qualified leads.* This is the most common and most worthwhile objective of an exhibition. We must ensure that the leads are qualified and prioritized.

3.   *To create, enhance and maintain the company's image.* From a
        negative viewpoint, our absence might be regarded as a bad sign.
        However, we should take a much more positive approach and use
        the exhibition to develop our image, reassure existing customers and
        attract new customers.

4.   *To obtain editorial comment.* Exhibitions are covered by a vast press
        machine, and editors are hungry for good material. Copy should be
        carefully written so that it will claim both the editors' and the
        readers' attention in the midst of a mass of conflicting messages.

5.   *To carry out market research.* This is discussed in 'advantages'
        above.

6.   *To assess the competition.* At most exhibitions, all the main competi-
        tors will be present. We should take the opportunity to observe what
        the competition is saying and displaying (or not displaying).

7.   *To support and find agents or partners.* It is common for an agency
        agreement to require both parties to participate jointly in the most
        relevant exhibitions, on a combined or adjacent stand. Exhibitors
        may also be able to find new partners at an exhibition. Whether they
        like it or not, they will be approached by various people promising
        them the earth! Agreements signed up in the euphoria of an exhibi-
        tion can prove to be disastrous, and thorough checking is necessary.

TARGETS

Having established the overall company objective, it is important that individuals are clear about what is expected from each of them. In many cases, a
special incentive should be given for business generated at the exhibition.

Targets might be based on the codes listed under 'Follow-up' below. There
are three main reasons for having these targets:

1.   To motivate and to judge the performance of individuals and the
        whole team.

2.   To assess the progress of an exhibition and take corrective action if
        necessary. We might decide after the first day that we need more
        staff or a different type of staff, or that we should rearrange the stand
        in some way.

3.   To help us to decide whether the exhibition was worthwhile and

whether it is worth taking a stand next time. If no criteria for success were established in the first place, we cannot judge whether they were met.

INVITATIONS TO EXHIBITIONS

Exhibitors should not rely solely on the invitations sent out by the exhibition organizers. They should have their own detailed lists of actual and potential customers. Names, titles etc. must be carefully checked. Possible groups include:

1.   Existing clients invited centrally, by the Marketing Department or senior management.
2.   Existing clients invited by individuals, for example by senior managers or by technical specialists.
3.   Potential clients known by someone in the company.
4.   Potential clients not known by anyone in the company.
5.   Existing clients of other parts of the organization who might be potential clients for the company which is exhibiting.
6.   Potential clients known to be working with the competition.
7.   Others.

Very differently worded invitations should be sent to each of the groups listed above (but they rarely are).

THE RIGHT OPENING QUESTION

'Can I help you?' This is probably the worst opening question. It invites the answer 'no thanks – I'm just looking', or 'yes please – I'd like to hear about your service in detail'. This can close the door on some potentially good prospects who do not want to reveal their interest. Conversely, it can open the way to time-wasting conversations with people who have no genuine intention or authority to buy.

A non-threatening question such as 'what use do you make of...?', 'what sort of problems do you have with...?', 'would you like to be able to save money on...?', 'what are you hoping to find at the exhibition?' is much more likely to encourage the prospect to talk in the way we want.

FOLLOW-UP

There should be some simple means of recording details of visitors, and all personnel staffing the stand should use the same system so that follow-up can be arranged on a consistent basis. The name, title, address and job title can most easily be recorded by stapling a visiting card to the enquiry form. The most important information is then the follow-up action. We must decide and record who is going to do what by when, before we speak to the next visitor. Scanning a bar code does not constitute a qualified lead.

A simple code can be used to define categories for follow up such as:

1.   A visit arranged on the spot.
2.   Visitor to be telephoned within one week to arrange a visit.
3.   Specific information to be sent on . . . . and then followed up.
4.   Contacts to be made after . . . . . weeks to arrange . . . .
5.   No immediate interest, but keep on mailing list.

Other information would be useful but is not always obtainable. The enquiry form might have a checklist to be used as appropriate. Examples are:

O   What is the contact's role in the buying decision?
O   Who else is likely to be involved?
O   Has there been any other contact between the two companies?
O   Is there a budget?
O   When is a purchasing decision likely to be made?
O   What is the visitor's specific interest?
O   What details of the current operation are relevant?

MAKING THE EXHIBIT MORE SUCCESSFUL AT ATTRACTING VISITORS

We should find a way of differentiating our stand from all the others in order to attract visitors. Exhibits increasingly tend to look the same. Computer displays are often unintelligible, and some equipment is not particularly attractive. We need to make imaginative use of layout, equipment, scale models, demonstrations, sound, light, display panels and staff. The biggest draw at a zoo is the feeding time for the penguins, and a little ingenuity may suggest some relevant and regular attraction which we can announce in advance and which will draw people to the stand. As an example, a company involved in industrial safety arranged to have a dust explosion every 15 minutes.

We finish where we began. Exhibitions can be the best and the worst way of spending marketing money. Some people who have never considered taking a stand might be pleasantly surprised by the results if they followed the above guidelines (Follow-up 9.4).

9.7   ADVERTISING

As we know from our daily experience, there is a very large industry dedicated to the business of advertising. Two parties are involved when we advertise, apart from ourselves and the target audience. First, there are the communications experts such as agencies, designers, media specialists, copywriters and many others. Their role is to create the campaign and the message, and to implement it after agreement with the client. Second, there are the media – the vehicles for the message – which include television, radio, cinemas, newspapers, journals, magazines, posters and point-of-sale material, as well as the more exotic examples such as racing cars, lasers, sky-writing, balloons, taxis, supermarket trolleys, toilet doors and who knows what next!

There are two distinct aspects of advertising – the *message* and the means by which the message is *communicated* (see Section 7.4).

The creative and logistic sides of communication are specialist functions for which it may well be worth employing an expert. Readers who wish to study advertising in more detail are recommended to read a book on the subject. This will give advice on matters such as the choice of an agency, developing an advertising campaign, selecting the media and assessing the results of the campaign.

As advertisers, the following issues are important to us:

1. We cannot assume that people will spend time reading our advertisements. We probably have about two seconds to capture the initial attention of the reader or viewer. The consumer marketing purist would argue that we can say only one thing in an advertisement, and that this must be said very succinctly and with high impact. In a high technology environment the message can be more complex than this, but it is healthy to bear the principle in mind. The instinct of many people, particularly scientists and engineers, is to say far too much, with the result that very little actually gets through.

2. We should bear in mind the progressive communications process described in Section 7.1. What is the purpose of the advertisement? If it is not to sell, what *is* it there to do? Is it to sell the next step? What is the next step? How quantified are the objectives? How do we judge its success?

Advertisers with large budgets would submit their advertisements to a panel, perhaps using modern psychological and other techniques to judge the potential effectiveness of the advertising. The cost of 'space', whether in a journal, a newspaper, on television or some other media, can be considerably higher than the cost of originating the advertisement. It makes sense to test the advertisement, probably in several different versions, before irretrievably committing the money to the media.

If our budget does not justify this, we might test an advertisement on a panel that we have set up ourselves. The panel should preferably be composed of potential customers who have not bought from us in the past, rather than loyal customers or members of our own company who will have a biased reaction.

An alternative to print advertising, i.e. space on a page of an appropriate journal, is a loose insert. This has four advantages:

1.   It is more likely to be seen – at least the recipient has to make some effort to throw it into the bin!

2.   Even if it is not read straight away, it may be kept in a pocket, file or drawer from which it can emerge at a later date.

3.   It can easily be passed to a colleague for whom it is more appropriate. Admittedly the insert is following most of the other ways of advertising into disrepute through over-use, but the better journals limit the number of inserts in any one issue.

4.   Many journals allow selective insertion, splitting readers by location (e.g. county), job title or in some other way. This means that expenditure can be limited to the main target audiences, and it also offers an opportunity for a restricted 'test market' before investing in full coverage.

A careful comparison of the relative cost of inserts and print advertisements needs to be made, together with an assessment of their response rate; the rest is simple arithmetic.

It is useful to apply to advertising the test with which this book opened. If marketing is not the means of generating profit, there is something wrong. My own seminar business has been built up by means of hundreds of thousands of inserts over a ten-year period, much of which has been in a recession. The advertising from the earlier years is still bearing fruit today, and the whole campaign has yielded an excellent profit (Exercise 9.6).

9.8   PUBLIC RELATIONS

GENERAL

To some people, PR and marketing are almost synonymous. This view reveals a very limited understanding of marketing. It also under-values the specialized role which PR has to play in the marketing communication process.

Public relations is one of the means by which the 'personality' of a company or service is created and communicated. In terms of Figure 7.1 in

Section 7.3, PR is an important but relatively low-key activity which starts long before the buying process commences and continues for an indefinite period afterwards. Its purpose is not primarily to sell or even to communicate detailed messages about the services. It is concerned more with enhancing the reputation of the organization. If, when the name of a company is mentioned, our mind spontaneously turns to reputability, professionalism, quality, reliability or high standing in the marketplace, it is probably because we have been influenced by an effective PR process, perhaps over a period of many years.

As with all forms of marketing communication, PR can be carried out by internal staff, external consultants or a combination of both. The role of the external consultant is to advise, create, challenge and, if required, to manage the PR process and key events in the programme.

PUBLIC RELATIONS ACTIVITIES

The type of PR events that are most appropriate varies enormously with the industry. In a high technology field, activities such as articles in scientific publications, papers at conferences, educational aids, panels, open days or the sponsorship of a university Chair might be most appropriate. For a service appealing to the general public, television, radio, visits and the sponsorship of sporting, cultural and other events might be used. For a company with a local marketplace, some newsworthy gift or activity might be cost-effective. For all of them, activities such as influencing VIPs, politicians and city opinion formers, press releases, press conferences and newsletters might play a role.

PRESS RELEASES

Press releases are considered separately because many companies produce them without drawing upon external expertise. The following guidelines may be helpful.

1.   *Have something which will capture people's attention.* This may be a good headline, an interesting human story, or a really creative use of photography or graphics. People do not read every word, and boring technical details do not arouse the casual reader's interest.

2.   *Realize that editors are hungry for good copy.* If we are known to provide relevant material that requires the minimum of amendment, they are more likely to be receptive to our offerings in the future.

3.   *Assume that the editor owns a pair of scissors!* We want as much

space as possible; the editor has a certain number of column centi-metres to fill. It is very annoying if the heart of our message is cut out for lack of space. The main points should appear at the start, and we should assume that our submission will not be printed in its entirety.

4.   *Take advantage of the fact that an article may have more credibility than an advertisement* because it is seen as being more objective and independent. Furthermore, it costs nothing.

5.   *Accept that, unless we maintain a very strong hold over the media, a great deal of luck is involved.* A brilliant article might be rejected because that particular issue is already full. We may need to submit many times to obtain the coverage we want.

6.   *Remember the key questions* – 'what are we trying to say?', 'to whom?', 'with what objective?' This applies as much to PR as to other forms of communication (Exercise 9.7).

CORPORATE IDENTITY

We have discussed the fact that the image of a service can be a key factor in the buying decision (Section 2.3). The image of the company which supplies the service, or the 'corporate identity', may have an equally strong influence. Rightly or wrongly, people will make assumptions about the service on the basis of the 'stable' from which it comes (Section 2.9).

A corporate identity programme aims to exploit this fact. It develops, communicates and reinforces the identity, reputation and personality that the company wants the market to recognize. An important aspect of such a programme is that the same image is presented across the whole range of communication media – not only the more direct forms of marketing communication such as advertisements and brochures but also items such as letter-heads, envelopes, signs on buildings, vehicles, ties and anything else that can aid recognition.

BROADER ISSUES FOR PUBLIC RELATIONS

Public relations can achieve objectives which are less direct than the promotion of the company or its services to the customers. We may wish to appeal to the community by showing that we are good citizens or employers, or that we care for the environment. We might want to show that we are concerned about future generations. We might want to be philanthropic by sponsoring worthwhile activities such as health or education. It is very difficult to attribute a specific sales volume to such actions, but they can reinforce and give credibility to our more direct forms of marketing (Exercise 9.7).

EXERCISES

9.1 Evaluate a piece of your company's promotional literature against the 12 guidelines in Section 9.1, awarding marks out of 5 for each factor. What would you do to improve the literature next time it is printed?

9.2 Do you have a problem with multi-purpose literature or with product-oriented literature? How could you create a cost-effective hierarchy of market-oriented literature that would serve different market segments more specifically?

9.3 Select a recently submitted proposal for a contract that you might have expected to win but which was *not* awarded to you. Evaluate it against the 12 guidelines in Section 9.2. How could you improve the proposal if you had the opportunity of resubmitting it?

9.4 Look at your web site in the light of the criteria suggested for promotional literature in Section 9.1, with the additional challenge about the accessibility of the different web pages to the serious seeker or the random browser.

9.5 Take from your filing cabinet or hard disk a letter which you have recently sent to a potential customer but which did not achieve the desired response. Evaluate it against the guidelines in Section 9.4. How could you improve the letter if you had the opportunity to rewrite it?

9.6 Ask a colleague to select an advertisement from a company in a field similar to your own. Look at it for two seconds and then write down what you can remember about it. If you were a potential customer, would you read any further? Having read it, would you take any action? Repeat with other examples. Apply the same approach to your own advertising. How could you improve its effectiveness?

9.7 Consider the PR activities currently used by your organization. Which of them are not as effective as they should be? Are you missing some opportunities?

FOLLOW-UP

9.1   Cultivate the habit of writing down the key 'win themes' before beginning to write a proposal. Evaluate it critically against these themes before submitting it.

9.2   Consider whether a reply form of the type discussed in Section 9.4 could be appropriate to any part of your business. If so, design a form, test it on a sample of potential customers and evaluate the results.

9.3   Make a summary of the points on using the telephone in Section 9.5 that are relevant to your business. Have this on display by your telephone to use as a checklist for the next three months. Modify the list in the light of your experience.

9.4   Before staffing your next exhibition stand, acquaint all the staff involved with the content of Section 9.6. If possible, hire a training video or arrange an exhibition workshop with a suitably experienced consultant.

PART III
THE ROLE OF BUSINESS DEVELOPMENT

❖

10

WHAT DOES THE MARKET WANT?
MARKET RESEARCH, MARKETING RESEARCH

Key business issues   *Section*

O   Business decisions depend on knowledge of the market.
    The cost of acquiring the information increases rapidly
    with the accuracy sought   10.1

O   We should be 'disciplined entrepreneurs'

O   Market research information is divided into 'primary' and
    'secondary' data

O   Market research can be qualitative or quantitative

O   Interviews and group discussions can be used very
    effectively for qualitative research if certain principles
    are adhered to and risks avoided   10.2

O   Existing in-house market intelligence is often
    under-used because it cannot be accessed

○ Quantitative research based on small samples should be left to the experts, but is a good investment in the right circumstances   10.3

○ Some types of market research do not require a large budget   10.4

○ Alternative options for marketing our services should be researched in order to evaluate the comparative effectiveness of each   10.5

10.1 BUSINESS DECISIONS DEPEND UPON KNOWLEDGE OF THE MARKET

The whole emphasis of this book is that we start with the needs of the marketplace and work back to our own resources (out-to-in). While this is partly an attitude of mind, market research is obviously an important ingredient of this approach. In some cases it can provide us with information on which to base our entire business strategy.

A problem with which we are immediately faced is that the cost of the research rises rapidly with the precision of the information that we wish to obtain. A relatively small sample and a low level of research activity can yield some apparently useful results, but there is a danger that the sample is unrepresentative and that the conclusions are therefore suspect.

A guiding principle is suggested. With a particular level of understanding of the marketplace, we are able to make business decisions with a certain degree of confidence. The question is 'would these business decisions be significantly better if we had more precise information?' If so, we should perhaps go at least to the next step. If not, why spend any more money? We cannot afford to pay for information just because it is interesting.

The reader is cautioned against the attitude that appears to prevail in some companies – that no business decision will be made until it is thoroughly ratified by research. This may be perfectly appropriate in large fast-moving consumer goods markets, where the rewards for success and the penalties for failure are enormous. In other cases, it may be an excuse for not making decisions. Management is the art of weighing up imponderables; we often have to make decisions when we have only half of the information we would wish to have.

This raises the issue of management style, as illustrated on the spectrum shown in Figure 10.1. The question for each of us is 'where should we be on that line?' Again, it depends upon the rewards for success and the penalties for failure. If we are designing power stations, we will be at the right-hand end. If we are working in a rapidly changing market such as software, there will be no

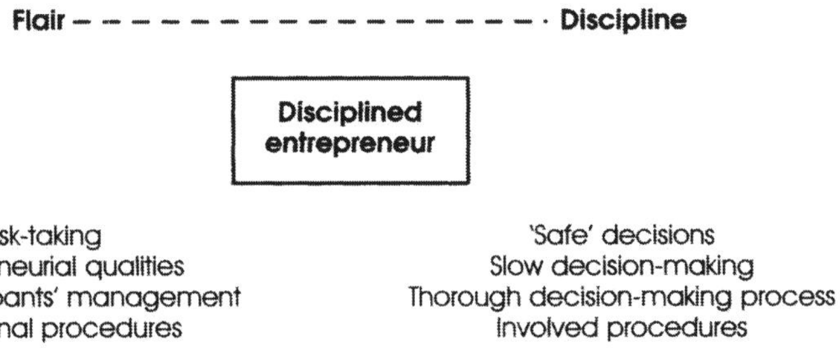

Flair – – – – – – – – – – – – – – – · **Discipline**

```
┌─────────────────┐
│   Disciplined   │
│  entrepreneur   │
└─────────────────┘
```

Risk-taking   'Safe' decisions
Entrepreneurial qualities   Slow decision-making
'Seat of the pants' management   Thorough decision-making process
Few formal procedures   Involved procedures

Figure 10.1   Management style

time for long drawn out decisions and we will be at the left-hand end. One seminar delegate said to me 'our company is about two yards to the right of "discipline" – we never make mistakes, but we never make any progress either'! The answer for many of us is to be 'disciplined entrepreneurs': we do whatever research is appropriate, give a sensible amount of management time to the decision, and then back our judgement (Exercise 10.1).

PRIMARY AND SECONDARY DATA

There are two types of data about the marketplace – primary and secondary data. These can relate to both customers and competition.

Secondary data is information that is already potentially available. The process of acquiring it is described as 'desk research'. It might be found in our own internal records of sales and customers, or it may be sought from external sources such as government statistics, industry and trade journals, economic studies, libraries, professional and academic reports, directories and so on. It takes time to search out the data, and it is not always in the form we would most like. Nevertheless, the cost compared with studies that we have specially commissioned is trivial, and it is irresponsible not to make use of it. A key to the effective ongoing use of secondary data is to design a suitable acquisition, storage and retrieval system (see Section 10.2). This can be supplemented by the use of on-line subscription databases if appropriate.

Primary data is information that results from new research studies which we ourselves have commissioned. It may be possible to reduce the cost by commissioning a 'syndicated' or shared project with other suppliers, but this has the disadvantage that the information is not exclusive to us and may not be in the form which we would ideally have chosen.

A confusing factor is that secondary data should be considered first! It is much less glamorous to sift through secondary data than to commission a new primary research project, but the cost is also much less. We should therefore start with secondary data, and commission a more expensive primary research study only if the secondary data is not adequate.

METHODS OF SAMPLING

If the number of potential customers is small – if we are selling a service such as computer integrated manufacture to car manufacturers, for example – it may be possible to carry out a census, i.e. to gain a knowledge of the whole marketplace. In most cases, however, we have to study a small sample and use the results to represent the whole. There are two methods of doing this:

1. A random sample. To avoid unacceptable risk, the sample must be statistically large enough and must have been obtained in an unbiased way. A random sample is not the first ten or hundred people we happen to meet. Nor is it likely to be found within our company, where our colleagues may be totally unrepresentative of the market both in buying power and in their attitude towards our service.

2. A structured quota sample. In this case the structure of the whole marketplace is carefully analysed and particular groups are differentiated by size, buying patterns, use of outside resources or other parameters. A specific number of interviews are then held within each group, and the significance of the feedback is statistically weighted according to the number of people that each sample is seeking to represent.

QUALITATIVE AND QUANTITATIVE MARKET RESEARCH

Market research activities can be divided into two types – qualitative and quantitative. In the first we are looking for information about things such as the desires, attitudes and behaviour of potential customers; in the second we are quantifying specific aspects of the information, in terms of either numbers or percentages.

10.2 QUALITATIVE MARKET RESEARCH

POSSIBLE OBJECTIVES

Opportunities for new services

In most businesses the cost of developing a new service is high, and it makes good business sense to evaluate the prospects as far as possible before investing in the service itself. The feedback may show that the service will be well received, or it may indicate some areas for improvement. Sometimes the research will convince us that we should not proceed with that particular service, perhaps because the competition is already dominating the market or because there is no money available for this particular service. Although this will be disappointing, it is extremely valuable information because it avoids further abortive investment.

Characteristics of buyers and users

Important input to our marketing strategy and tactics can be obtained if we study issues such as:

O   the way in which the service is to be used
O   the issues which surround the buying decision
O   the role of different members of the decision-making group in the buying process.

Image and perception

Image research should not be dismissed as being of no value just because image is intangible. As we have discussed in Chapter 2, the perception of our service, our brand, our company or even ourselves as people can override all other considerations in the buying decision.

I once saw an image survey that included a question on the installation lead time of sophisticated electronic equipment. The rational part of us says 'there is no point in carrying out image research on installation time – that is a fact not an opinion'. Wrong! The key issue is not how long it will take, but how long people *think* it will take. If their perception is that one company will take twice as long as another, they will react accordingly. This is true even if their perception is wrong, or if it is out of date. In this particular case, there had been a problem with one company but they had put it right; the survey showed that the perception remained and that they were losing business as a result. Thus, if people think we are slow, bureaucratic and expensive, even if they are wrong, their perception will influence the buying decision.

Market segmentation

As discussed in Chapter 4, our services and our marketing messages need to be targeted at particular segments. It may be necessary to do some market research to enable us to segment the market in the most effective manner, in order to decide the most appropriate marketing approach to each.

METHODS

Individual interviews

The most effective form of qualitative research is a discussion with a sample of respondents, probably in the interviewee's office. This will take the form of a 'depth' interview, using topic guides as prompts to open-ended questions to encourage the respondent into talking in the desired direction.

The process may be carried out on a single occasion in order to obtain a 'snapshot', or it may be repeated after a period of time in order to observe trends during the intervening period. Careful structuring of the questions is essential. Ambiguity must be avoided because this would invalidate the whole study. The respondents must feel that the questions are relevant or they will lose interest and either terminate the interview or give their answers less thoughtful consideration.

The questions must be asked on a uniformly structured basis so that the answers can be correlated and consolidated. The safest route is perhaps to use a well-briefed consultancy for the whole project; the advantage of this is that they are specialists in the area and are aware of the pitfalls. However, it is my experience that with carefully worded questions one can obtain at least some valuable information at a much lower cost (see Section 10.4).

Group discussions

Sessions may be run by consultants using 'focus groups' or other formal techniques. In this case, one is paying for specialized experience and facilities. Alternatively, it can take the form of a discussion panel or 'brainstorming' session. Again, with certain safeguards, we can carry out such research ourselves on a relatively low budget as discussed in Section 10.4.

It has been cynically suggested that focus groups never come up with anything innovative. If this is so, it is because the programme has been incorrectly structured rather than because the technique itself is flawed.

In-house information

There is one source of market information that is 'under our nose'. It is to be found in our sales and invoicing records, our customer correspondence files, the minds and reports of our staff, and in various other places. The problem is that we cannot access this information easily or draw statistically significant conclusions from it if we have not first set up some system to categorize and analyse it. We should also be aware that people in a sales situation tend to react disproportionately to the last thing that happened; we may need to receive similar reports from a number of sources before we attribute great significance to the information.

A little effort to design a suitable system can provide useful results. Someone needs to be given the responsibility of maintaining the system, and everyone who receives information from the marketplace must conscientiously pass it on to be entered on the system according to the agreed guidelines.

The system then provides a structured database of customers who can be approached under various circumstances. We might ask their opinion on a proposed new service, or visit a sample of them to discuss their longer-term needs.

The same approach can be used to gain an understanding of our main competitors.

RISKS

Before any form of market research is undertaken the risks must be very carefully evaluated. Without this, the results can be not only useless but misleading. Those setting up and carrying out the exercise must clearly understand issues such as bias, sample size and statistical significance. The main risks are:

1.   *The group is unrepresentative.* Methods of obtaining samples have been mentioned in Section 10.1.

2.   *The interviewer influences the discussion.* As with scientific research, the essence of all market research is that it must be totally objective. It is not only a waste of time and money but it is actually a delusion to conduct research which simply confirms our prejudices. If there is bad news, we need to know it. The danger is that interviewees or members of a panel tell the interviewer what they think he or she wants to hear. The skill is to devise appropriate questions and to probe in an unprejudiced manner so that we learn what the respondents really think.

3.   *The analysis and the report are biased.* This is a subtle trap. Some people pay consultants to tell them things they already know and want to hear, and the consultant is anxious to produce a report which will be welcomed (am I becoming cynical?). It can also arise when the study is carried out by internal staff who want to reinforce their own views or impress their superiors. Ruthless objectivity is essential and, indeed, some of the most valuable research comes when we find information which is, at least at first sight, surprising or even unwelcome (Exercise 10.2).

10.3 QUANTITATIVE MARKET RESEARCH

Quantitative research is used either separately or in conjunction with qualitative research when we want to put numbers on the information. As we have said, the cost of the research rises rapidly with the precision of the information we wish to obtain. With the exception of research on individual customers, discussed in Section 10.4, the field of quantitative research should be left to specialists.

POSSIBLE OBJECTIVES

Common objectives are to obtain figures on market size, market shares and trends. If the marketplace is at all sophisticated, the amount of information required, and therefore the cost of obtaining it, can be extremely large and, so, we may have to be selective. If we have a broad idea of the market size and the share of ourselves and our main competitors, the most useful information is probably the trends. Uncertainty in market size and share may not significantly affect our business decision making, but it may be essential to our future investment plans to know whether the whole market or the market share of a key competitor is increasing or decreasing by a certain percentage each year.

The classic users of this type of research are the large consumer marketing companies. The marketplace is almost treated as a computer model on which certain changes can be imposed, so that the effects of a new product launch, a price change, a promotion, an advertising campaign and so on can be measured. Past activities can be evaluated and decisions made about future programmes. The cost of this regular research is very high, but it makes eminent sense in this type of situation in view of the very high rewards for success and the penalties of failure. For many high technology readers, this type of market research would not be appropriate.

METHODS

The market research information described above can come from regular (usually monthly) audits, or it may be obtained on a 'one off' basis, possibly repeated after a period of time in order to measure a trend. The methods include recording the levels of business, face-to-face interviews, or telephone or postal questionnaires.

RISKS

The risk of a wrongly structured sample is so great that the task should not be undertaken by those with no specific expertise in this area. Having set up the sample, the target list of respondents must be rigorously pursued. If they or a legitimate equivalent are not found, the whole exercise may be biased (Exercise 10.3).

10.4 MARKET RESEARCH ON A LOW BUDGET

'The easiest answer to all market research problems is simply to commission a consultancy with the appropriate expertise, pay the fee and all our problems will be solved!' Apart from the fact that this is not true, it is not very helpful advice for companies with little or no money available to spend on professional market research.

While there are some areas of research that should only be performed by professionals, there are four approaches which I have used successfully (and I would not claim to be an expert in the specialist field of market research). These are a group of opinion formers, users' groups, image research and quantitative research on individual customers.

A GROUP OF OPINION FORMERS

Assembling a suitable group is a useful means of gaining an informed view of the longer-term trends in a particular industry. The group might typically consist of between three and six people whose opinion can be trusted and who have played a significant part in the development of the industry.

The group might include leading academics and research workers (as long as they have their feet firmly in the commercial world), a senior representative of a professional or other body if appropriate, an informed journalist or other commentator on the scene, and some senior representative users or recently retired users who are not prejudiced or restricted by commercial confidence.

The key to success in such an event is to have a really effective moderator, facilitator or chairperson. The first task is to attract the members of the group. I have seen senior people willing to give up two or three hours, without payment of a fee apart from their expenses, simply because of their respect for the person leading the session. The chairperson should be well aware of the objective of the session, so that the right balance can be struck between allowing freedom of expression and wandering off into unproductive side-tracks. The emphasis is likely to be on future business needs rather than on the technical routes to meeting these needs.

Some consultants would record or video the discussion, but I personally prefer to do without such intrusions. The session must be unthreatening so that people are free to express their views without any inhibitions. One or two senior members of the company might be present, unobtrusively taking notes and participating from time to time but not in any way dominating the discussion. After the event they then quickly 'capture' all the main thoughts which have been developed, and record them for further consideration.

As with all market research, it is essential that the participants do not simply say what they feel the company would like to hear. It must not be regarded as a political or selling event, and the views expressed must be as objective as possible.

USERS' GROUPS

Users' groups are somewhat similar to the above, but the members of the group are all existing users of the company's services. They meet from time to time in order to advise the company on future lines they might follow. In terms of time-scale, whereas the first group is looking very much to the long-term speculative future, a users' group is looking at more immediate issues such as an improved version of the service they are already using. It is therefore more pragmatic and possibly more technical than the first type of discussion.

Again, it is important that the meeting should not be seen as a commercial or selling event. It must not become a forum for criticism. If a member of the group is likely to complain about the company's performance, legitimately or otherwise, the issue should be dealt with outside the discussion; if it is not, the complainant will act as a 'rotten apple in a barrel', and the whole event will be counter-productive.

In my experience, users are often happy to host such events on their own site and, indeed, are quite proud to show their facilities to non-competing visitors. Again the cost to the company is trivial – perhaps some modest refreshments and expenses. Compared with the cost of commissioned research, the cost–benefit ratio is extremely favourable.

IMAGE RESEARCH

We have stressed the importance of knowing how we are regarded by the marketplace. There are obviously times when a large externally commissioned image study is worth while, but there may also be times when a relatively modest exercise can yield useful results.

One such example is a high technology consultancy organization that had had a disappointingly low success rate with the award of contracts. They decided to visit a sample of the potential clients to find out the reasons for rejection. The feedback from the first four respondents was unanimous and therefore highly statistically significant. The problem was not the price or any doubts about technical ability. It was a perception that the people who would carry out the projects were not sufficiently attuned to the highly competitive commercial environment in which the companies had to operate. This perception was not accurate, but that is not the point – images do not have to be true to affect buying decisions positively or negatively. The logical outcome of the research was therefore to find the cause of the image problem. This is an important lesson which many organizations need to learn.

It might sometimes be useful to carry out a pilot study in order to decide whether or not to invest in a larger research project.

QUANTITATIVE RESEARCH ON INDIVIDUAL CUSTOMERS

As with all market research, the study must not be regarded as 'selling in disguise'. Unless the responses are objective and free from bias, the exercise is not only useless but may actually be misleading, quite apart from the fact that this is not a proper use of market research.

The aim is to find out specific data on individual customers. Let us suppose that we are a leading consultancy selling high-value services relating to computer installations used in industry. We might choose a representative sample of existing users of our own and competitive services. We would then visit them with a questionnaire, asking for information such as:

O   When are you proposing to install a new system?
O   For what reasons will you be wanting a new system? (Higher capacity, new functionalities, improved flexibility etc.)
O   Within what price bracket is the new system likely to lie?
O   How much outside consultancy input are you likely to need?

The quantitative aspect of this research comes when the results of individual interviews are combined in a way that will provide a numerical input to future product strategy.

The risk of this type of research is that it costs the respondent nothing to be optimistic. If asked 'would you like a service that can increase the capacity of your present equipment?' most people would say 'yes'. To persuade them some time later to pay for the service is a totally different matter! The questions must therefore be very carefully posed, and there is an advantage in using an external interviewer who is not seen to belong to one of the supplying companies. The information received can be extremely valuable compared with the cost of achieving it. As with other forms of market research, the results of a pilot study might lead us to commission a more detailed study to probe crucial issues in more detail.

Where the potential target market is small, the research can take the form of a 'census' – i.e. covering the whole marketplace. In this case, there are obviously no problems of extrapolating from a small sample to the whole market, and bias of this particular type is eliminated (Exercise 10.4).

10.5   MARKETING RESEARCH

So far we have used the expression 'market research' to describe research into the market. This includes research into the competition.

However, there is another extremely important area of research which we might call 'marketing research'. This is researching and testing the various options for marketing our services. It involves carrying out experiments in the marketplace, measuring the results, drawing conclusions and thus obtaining an input to future decision-making. (It must be said that this distinction between market research and marketing research is not universally used; some people refer to the whole subject as 'marketing research'.)

Some of the activities that can be researched in this way include the following.

ADVERTISING

The degree to which it is worth researching our advertising depends very much upon the cost of the advertising. If we are planning to spend millions of pounds on a television campaign, it makes good sense to spend, if necessary, tens or even hundreds of thousands of pounds to attempt to ensure in advance that the ultimate investment will produce the maximum results. In these circumstances, our advertising agency might produce several versions of a proposed advertisement and test them with a panel, measuring such aspects as impact, recall, recognition of the company name, motivation to use the service and so on. The advertisement might then be tested in one television region before being released nationally.

On the other hand, if we have an advertising budget of only a few thousand pounds, marketing research might be confined to trying different forms of advertising and logging the responses against the date and place of the advertisement. This would help us to decide which journals to use, at which time of year to advertise, whether to use pages in the journal, its front or back covers, 'bingo cards' (where readers tick boxes for further information), inserts and so on.

In favour of this latter more pragmatic approach, it must be said that the only real test of an advertisement is whether it works. Impact, recall and so on are not ultimate objectives of advertising, and they do not always correlate as precisely as we would hope with the level of orders or enquiries that they generate. Advertising which wins prizes at film festivals does not always do the best job for the advertiser.

SELLING

Again, because a large amount of money is involved in the selling process, it may be worth researching the effectiveness of different methods of selling and various ways of combining them. Options might include:

1.   *Specialist versus generalist sales people.* For a newly introduced service, there might be an advantage in dedicating part of the sales team full time to exploiting the opportunities for an initial period.

The rest of the sales force might be asked to pass on sales leads to the specialists during this phase. Some might say that the commission arrangements would not permit this, but who is making the rules?

2.   *Full-time versus part-time sales staff.* Many high technology companies, such as those involved in consultancy, have to use fee-earning staff for the selling process, because they have no one else. This has advantages and disadvantages, as discussed in Section 8.6. These options might form the subject of a marketing research study.

3.   *Selling direct versus selling through agents.* This a more fundamental decision, in which we are testing whether the cost of the fees and commission allowed to the agent is more or less favourable than the cost of selling direct to the end customer through our own resources. If we are considering a change, a pilot experiment might be set up as part of a marketing research exercise.

LITERATURE

Different versions of a particular item of literature can be tested at the draft stage before committing to the final version. Alternatively, we might test the effectiveness of having a whole hierarchy of literature with a specific purpose for each component. This could involve targeting different market segments or different members of the decision-making group, as discussed in Chapter 7.

EXHIBITIONS

I once took an exhibition stand where the cost had been recovered by 10.35 a.m. on the Monday of the week following the exhibition. This is a rather precise measurement that is not available to everyone! Unless we have a specific and measurable objective, it is impossible to determine whether an exhibition was worthwhile and whether we should attend on the next occasion.

Marketing research might therefore test one exhibition against another, and could test the relative effectiveness of spending money on exhibitions compared with other ways of spending the same amount of money.

PRICE

As price can have such a significant effect on the volume of sales and their profitability, any research which will assist us in determining the optimum market-based price can be very valuable.

As discussed in Chapter 6, many companies under-price in some areas and over-price in others, because they do not know the price the market will bear. In a large marketing operation, where a reasonably standard service is being sold in large quantities, price can actually be tested by selling at different prices in different regions of the country (assuming all other factors are equal). In a consultancy situation, where the price is decided for each individual project, the broad effect of increasing or reducing prices can be judged.

TEST MARKETS

A highly sophisticated form of marketing research is the test market, where the entire package involved in the launch of a service is tested on a limited scale before committing to a national or international launch. This approach offers great benefits, yet it is not even considered by many high technology service organizations (see Section 11.4).

Marketing research is almost an attitude of mind. It means that we recognize that we may not yet be marketing our services in the optimum way, and that there is rarely one 'right' way to market a particular service, so we are constantly trying new approaches and learning from our experience. As with normal market research, the cost need not be very high if we use our existing resources with imagination and intelligence (Exercise 10.5), (Follow-up 10.1).

EXERCISES

10.1 Where do you think your company lies on the spectrum in Figure 10.1 in Section 10.1? Is this appropriate for your type of business? What changes need to be made?

10.2 What subject would you most like to research qualitatively in the marketplace? How would you go about it? What size of sample would be required? How much would you be willing to pay for such a study? What three (or more) main questions would you ask?

10.3 What subject would you most like to research quantitatively in the marketplace? Would you need to use external resources? With what precision would you like to know the result? How much would you be willing to pay for such a study?

10.4 What type of 'low-budget market research', as described in Section 10.4, might be appropriate to your operation? What subject will you research? Who will organize it? What risks will you specially consider? When are you going to do it?

10.5   Which parts of your operation would be useful subjects for marketing research? What alternatives will you want to test? Who will do it? How will it be done? When are you going to do it?

FOLLOW-UP

10.1   Cultivate the habit of asking yourself, every time you consider a particular form of marketing expenditure, whether you have sufficient evidence to say that this is the most cost-effective way of achieving the desired objective.

11

HOW DO WE ACHIEVE PROFITABLE INNOVATION?
MARKET-LED INNOVATION

Key business issues	Section
O Development of new services is not the only route to business development. New markets for existing services may give quicker returns, lower cost and less risk, even though the work is technically unexciting	11.1
O The marketing cost of innovation may exceed the technical cost and should be included in the appraisal process	11.2
O Differentiation should be built in at the concept stage	
O More new services fail for marketing reasons than for technical reasons	
O Proposals for new services or new markets should be put through policy, marketing, financial, management and technical filters before expenditure is approved	11.3

O Probable winners should be identified as early as
   possible. We should not 'start developing now, think
   about the market later'   11.4

O Income and expenditure forecasts for new services
   should be analysed for risk and sensitivity   11.5

O Delay in time to market can make a project
   unprofitable. Management attention is often directed
   at the wrong areas

O Impending failure should be recognized, and projects
   terminated before even more money is wasted   11.6

O Unwillingness to 'lose face' inside the company or at
   corporate level is a common cause of wasted
   investment

11.1 OPTIONS FOR BUSINESS DEVELOPMENT

Ansoff (1988) defines a diversification matrix which offers four possible
strategic options for business development.

We can sell new or existing services into new or existing markets, and so
the options are:

Penetration	Existing services into existing markets
Development of new services	New services into existing markets
Market development	Existing services into new markets
Diversification	New services into new markets

PENETRATION

Penetration is selling more of our existing services in our existing markets,
and is the routine daily activity of most businesses. Whether or not this is
sufficient to meet our corporate targets depends upon the size of the market,
its growth potential, our market share, the strength of the competition, the
relative merits of our services and so on. If the potential is sufficient, it may
be wise not to divert management time and money into new areas.

However, although such a favourable environment can exist, it may not
remain for long. If the market is sufficiently attractive, new competitors will
come in until potential supply exceeds potential demand, and everybody has
to engage in a constant battle to defend their own share. In these circumstances, something different has to be done.

DEVELOPMENT OF NEW SERVICES

This involves conceiving and developing new services for our existing markets. The first reaction of many managers when thinking about business development is to turn to the R&D Department. 'In order to grow, we must have new services', they say. This may well require research into new technologies that can form the basis of the new services. This may be the right thing to do, but it can be an expensive and risky process (see Section 11.5), and it is important to realize that this is not the only option open to us.

MARKET DEVELOPMENT

Market development involves selling our existing services into new markets. These may be new geographical markets, or they may be new market segments where a service that has been tried and tested in one segment is seen to have potential in another. The service may have to be adapted slightly to the new marketplace, but this route avoids the cost and delay attached to developing completely new services.

Experience with a large number of companies shows that market development is an increasingly common activity. In many cases, the market in which the company has been operating has either plateaued or is declining. Examples are the defence, nuclear and space industries as discussed in Chapter 4. The market will continue to have potential for evermore, but it is inadequate to satisfy all the participants. In these circumstances, companies have no option but to seek to sell their skills elsewhere.

Scientifically, this seems to offer no problem. However we need to consider the very substantial process described in Section 4.8 as 'establishing a track record'.

At certain stages in the evolution of a company, market development may bring greater and quicker rewards than new service development, and may involve less investment and less risk. The problem is that market development is not very exciting for the R&D Department. It takes them away from the stimulation of researching and developing new services, and requires them to undertake unchallenging tasks such as adapting the service to new standards. Some companies overcome this problem by assigning different staff with different motivations to the two very different operations.

In many cases, service development and market development are used in combination. A new service is due to appear in, say, two years' time, but in the meantime we have to maintain the profit stream; entering a new market with an existing service may be the best means of doing this.

FULL DIVERSIFICATION

Full diversification involves selling new services into new markets. It requires two steps into the unknown, and the combined risk may be very high indeed. There may be nothing wrong with having a small element of full diversification as part of a total portfolio, but it is very risky for this to be the only option remaining. Such a position should be anticipated many years in advance and corrective action taken. It is no use waiting until the demand for one's service has ceased, perhaps as part of the 'peace dividend', and then sitting back and blaming the government. Marketing is making the future happen, and this includes avoiding the unacceptable consequences of predictable future situations as well as exploiting the profitable ones (Exercise 11.1).

11.2 THE ROLE OF MARKETING IN INNOVATION

If you ask most companies how much they spend on innovation, they will quote the percentage of their turnover that is allocated to the R&D budget. This reveals a very inadequate view of what marketing is there to do. Many market-led companies would recognize that the marketing cost of innovation may have to be at least as high as the technical cost. In other words, new service development is not something that takes place solely or even mainly in the R&D Department. If this is so, we might feel that the marketing cost of innovation is so high that we cannot afford to engage in it. Successful market-eers would say the opposite. Without in any way playing down the importance of the technical content of innovation, the key to business success may lie as much in identifying and exploiting profitable market opportunities as in developing the services themselves.

Serious problems arise in large companies when the total cost of innovation is not taken into account when making investment decisions. Senior technical staff are sometimes authorized to decide on suitable R&D projects without reference to marketing. A senior technical director told me that projects were prioritized on a technical basis; this has to be wrong. One of the purposes of this book and the seminars on which it is based is to enable 'non-marketing managers' to understand the role of marketing, particularly in innovation, and to be able to interface synergistically with their marketing colleagues. The benefit of collaboration between strong marketing and strong technical staff has been discussed in Section 3.3 (Exercise 11.2), (Follow-up 11.1).

A key part of the specification of a new service is that it should be differentiated from competing services in the mind of the customer. The technical basis for this differentiation may be highly sophisticated – but it does not need to be. The important point is that we are launching a new service with some strong claims to uniqueness which will attract the customer. This differentiation should be built into the service at the concept stage.

More services fail to achieve commercial success for marketing reasons than for technical reasons. This needs to be recognized when investment decisions are being considered.

11.3 MAKE DECISIONS BEFORE SPENDING MONEY

It is sometimes easier to understand the technical implications of a new service than the marketing factors which are likely to be involved. The technical specification can be accurately defined, and the course of the project and the way in which it is managed can be expressed in a logical manner. Sophisticated techniques can be used during the project review procedure.

The marketing issues on the other hand are usually much less clearly defined, even in a market-led organization. The needs of the customer are not precisely known and the activities of the competition may be unpredictable.

For this reason, companies tend instinctively to start by spending their time and money on the areas with which they feel most comfortable. They embark on the technical part of a project in order to show some progress and to test the technical feasibility of the product concept. They advance with a large number of projects in parallel, in the hope that some of them will eventually achieve commercial success. The Marketing Department is brought in at the end of the process with the task of selling a service which it is hoped will meet market needs.

The saddest obituary I ever read in a development report was 'Project successfully completed, Marketing not interested'. This statement reveals a total misunderstanding of the role of marketing in innovation. Marketing do not have the option of 'being interested' once the project is complete. Their obligation is to commit themselves to marketing the service (subject to successful technical development) *before* any money is spent on R&D. The technical specification – the description of *how* the service requirements will be met – should not even be written until the Marketing Department has formally agreed at least the main parts of the user requirement or marketing specification – the definition of *what* the market will want the service to do. Where strong technical staff are interacting with weak marketing staff, they do not receive this authoritative and definitive input; they therefore go ahead with development, making 'marketing' decisions as they go, and hoping that the results will sell.

It is useful to consider a number of 'filters' through which all new ideas should pass before money is spent on them. This may seem to be a counsel of perfection, but why not? There are enough business problems that lie outside our control; we should at least try to resolve those within our control.

THE POLICY FILTER

O   Is there a positive policy to go into this area?
O   If not, does the policy allow the freedom to do so?
O   Is there a policy not to go into this area?
O   Is there a policy that another part of the company should do it?

THE MARKETING FILTER

O   Is there a market for the service?
O   Does it have money to spend in this area?
O   Is the market large enough?
O   Is it growing at a sufficient rate?
O   Do we have sufficient customer benefits to create a demand?
O   Do we have sufficient relevant differentiation to overcome the existing and future competition?
O   Do we have the calibre, type and quantity of marketing resources?
O   Do we have the calibre, type and quantity of selling resources?
O   Do we have the appropriate sales channels, or is there enough in the budget to set them up?

THE FINANCIAL FILTER

O   Have we completed a full financial appraisal (all areas of expense, including marketing, not only R&D)?
O   Are we able to commit the necessary financial resources – capital and revenue?
O   Is the predicted profit stream adequate?
O   Can we stand the initial profit and cash outflow?
O   Have we evaluated the risks and sensitivity? (See Section 11.5.)
O   Is the worst case acceptable?
O   Have we avoided the trap of self-fulfilling delusion? (See Section 11.5.)

THE MANAGEMENT FILTER

O   Do we have the necessary calibre and type of management skills?
O   Do we have the necessary quantity of these management skills?
O   Have we resolved conflicts of priority for these management resources?
O   If we are entering a new area of business, do we have the right style of management?
O   Do we sufficiently understand the new business?

O   Are we in danger of extrapolating too far from the known to the unknown in management terms?

THE TECHNICAL FILTER

O   Is the project technically feasible?

O   Do we have or can we acquire the necessary technical skills and resources at all levels?

O   Do we have the necessary spare capacity to devote to the project?

O   Have we resolved conflicts of priority for these technical resources?

The more we can apply these filters at an early stage in the decision-making process, the greater will be our chance of success and the less will be the amount of management time and money which is wasted.

The sequence set out above seems to represent a logical train of thought but, if possible, it would be best to identify and tackle the most difficult filters first. If the project then fails at the first filter we apply, at least we will have saved the management time involved in applying all of the others; if it passes, the rest of the process should be that much easier (Exercise 11.3), (Follow-up 11.2).

11.4 IDENTIFY WINNERS AS EARLY AS POSSIBLE

The process of choosing the projects in which to invest may go through a number of distinct stages. We are thinking particularly of a new service, but the same criteria apply to marketing activities such as entering new markets. Money spent on marketing decisions early in the process can reduce the amount spent on projects that are ultimately aborted.

PRODUCE A LONG LIST

Initially, it is useful to list as many ideas as possible. This can be done by using all likely sources of ideas within the company, perhaps by means of a brainstorming session. A marketing purist would say that all ideas have to come from the Marketing Department because that is their prerogative, but this is arrogance. All ideas should go through filters, including the marketing filter, but the number of ideas should not be limited at this stage. Sales people, R&D engineers, software specialists, senior managers, customers – indeed almost anyone – may have ideas worth considering. The art is to reduce the list as early as possible in the process in order to identify the relatively small number that should become committed projects.

SCREEN THE IDEAS

The initial screening of these ideas may reduce the long list to perhaps a half or a quarter of the original total. This screening may be done inside the company, but only if those making the screening decisions have a good understanding of present and future marketplace needs. The criteria must be based on market and financial considerations as well as technical.

The cost of screening each idea at this stage may be relatively low, possibly of the order of a few hundred pounds.

TEST THE CONCEPTS

The ideas which have survived the first stage now need to be tested, albeit on a theoretical basis, in the marketplace. This may be done using formal or informal panels, arranged by ourselves or by a market research consultant, or we may visit a number of key potential clients or opinion formers (see Chapter 10). The purpose is to judge whether the concept on which the possible new service is predicated is soundly based.

The cost of a concept test will be considerably higher than that of the initial idea screening. It is for this reason that we are proceeding in stages. Each stage covers a smaller number of services, and the cost of implementing each stage rises rapidly.

DEVELOP THE SERVICE

This part of the process represents the technical content of the innovation project. It is significant that this is the third item on the list and not the first. The cost of the project may vary from a few thousand to millions, tens of millions, or even hundreds of millions of pounds depending upon the nature of the project. A very little arithmetic shows the benefit of the first two preliminary stages. If the relatively small amount of money spent on idea screening and concept testing can improve the success rate of the R&D Department by even a small percentage, it will be amply rewarded.

CARRY OUT A TEST MARKET

Test markets are a standard procedure in the consumer marketing industry and in many others. The cost of a national launch can be millions of pounds; an international launch costs many times more. A small proportion of this amount of money spent on validating the programme before commitment makes eminent sense.

The test market will seek to test not only the acceptability of the service but also the effectiveness of pricing, advertising, literature and any other parts of the marketing programme which can be tested on a limited scale, i.e. on a small but representative part of the marketplace. This process should be contrasted with the relatively low-key efforts of many companies during the phase between development and launch of a service.

Various results of a test market are possible. If everything is favourable, we can launch with even greater confidence, using lessons learned during the test. Alternatively, the test market may reveal inadequacies in the service, the message, the media, the price and so on; if such feedback is potentially available, it is disastrous not to take steps to find it out (Exercise 11.4).

11.5   RISK AND SENSITIVITY

The normal starting point for any business plan and marketing plan is a sales forecast. The period of time that this covers depends upon the nature of the business, but a five-year period is common.

This sales forecast, which represents the planned turnover, is then set against all areas of expense such as the cost of fee-earners, R&D, marketing and general and administration expenses to give a net profit.

All these figures are, by their nature, estimates. A risk and sensitivity analysis asks 'how much does it matter if any one of these figures turns out to be wrong by a certain amount, and is there anything we can do to minimize the risk of this happening?'

In Table 11.1, the turnover has been based on an average fee for a service package of £5000. The Cost of Sales is then subtracted to give the Gross Margin in each of the five years. (Note that 'Cost of sales' is the cost of providing the service, assumed to be £3000 per unit in this case; it is nothing to do with selling expenses which appear lower down.) The gross margin is the amount of money which is available for two purposes – covering the indirect expenses not directly attributable to providing the service and, we hope, providing a profit. For simplicity in this example, the indirect expenses are expressed as a total figure. The brackets indicate that they are negative figures.

The total profit during the period is the sum of these annual figures, i.e. £525 000.

There is a danger that managers engage in the self-fulfilling delusion of putting in sales figures that give them the answer they want. A profit projection of this kind acquires a legitimacy and a sense of authority which many recipients are prepared to accept without question.

What can go wrong with such a forecast? The answer is – everything! The volume of contracts may not be achieved, the average unit selling price may become depressed by the activities of competition, the unit cost of sales may be exceeded by unforeseen increases in cost, and any area of expense may

Table 11.1

Year	1	2	3	4	5
Units sold	0	100	150	250	300
Financial (£'000)					
Sales @ £5000	0	500	750	1250	1500
Cost of sales @ £3000	0	300	450	750	900
Gross margin	0	200	300	500	600
Indirect expenses	100	200	225	250	300
Net profit	(100)	0	75	250	300

eventually turn out to be over budget. The question is – how serious are these variances?

It would not matter if 'swings equalled roundabouts', for example, if we failed to reach the unit price but exceeded the sales volume in a way which still gave the same bottom-line profit. While a purist would like the profit to be achieved in exactly the way planned, a realist will be thankful that it has been achieved at all.

The key to this exercise is, therefore, to evaluate very critically the risk of each figure being wrong by a certain amount, and the sensitivity of the profit to this variance. This can be done on a computer model, but it can also be done simply with a spread sheet or calculator. We can calculate the effect on profitability of a sales shortfall of, say, 10 per cent, 20 per cent, 30 per cent etc. Similarly, we can evaluate the effect of adverse variances in all the other factors. This very quickly shows where the sensitivity lies. In the example quoted above, the profit is completely eliminated if any one of the following occurs:

O   a price shortfall of 13 per cent
O   a delay of about 11 months
O   a sales shortfall of 33 per cent
O   a unit cost increase of 22 per cent
O   an expenses overrun of 50 per cent.

In reality, the variance will not be restricted to one factor but will be found in most if not all of the elements that contribute to profit.

Where do we go from here? We start by examining the areas where a variation is most likely to put the planned profit at risk, and then look for ways of containing this variation.

For example, in the case quoted above, a delay is disastrous. A little arithmetic will show whether it would be worth spending more money in the development phase – by bringing in subcontract staff, for example – to increase the probability of reaching the target time-window.

Further work to ratify the sales forecast and unit selling price would also be very worth while, although it is accepted that these figures will always be estimates. The figures are particularly important in cases such as this where most of the income is generated in the latter part of the period; this is, of course, the time for which the forecasts are least likely to be accurate.

The sensitivity of the profit to time to market has been estimated by calculating the effect of a one-year delay and then taking a proportion on a pro rata basis. Even this is too optimistic – it assumes that the competition does nothing during the intervening period and that the customers are happy to wait, neither of which is likely. The danger is that we become neurotic if the R&D budget is exceeded even by a small amount, but are not sufficiently concerned if the service misses the time-window for which it is intended.

Simple arithmetic shows that these priorities may be completely wrong. Having said this, good management will adhere to expense budgets as well as planned time-scales.

The above calculations contain a basic fallacy in that they have ignored the time value of money. If we use a 10 per cent interest rate, a pound today is equivalent to 75p in three years' time. The cash flow statement should be discounted, using the well-established DCF technique, to bring the expenditure and income back to present-day terms, in order to calculate a net present value (NPV). In most cases – and this is certainly true in the above example – we are spending the high-value pounds now and earning the lower value pounds a few years later. The effect of this is to bring the NPV of the project down to, perhaps, one-half or three-quarters of the arithmetical sum of the profit stream or even less, depending upon the discount rate used. A brief description of the DCF technique is given in the Appendix at the end of this chapter (p. 201).

The analysis may show that there is a possibility of favourable variances, so that our profit forecast will be exceeded. This may create problems, with availability of qualified staff for example, but let's have a few problems like that! (Follow-up 11.4.)

11.6 DEALING WITH IMPENDING FAILURE

We have been arguing the case for a thorough investigation of marketplace needs and an analysis of the competition to arrive at a sound marketing specification. We have also urged that a full financial appraisal of the investment is made before substantial funds are committed to research or development work.

In spite of all these precautions, it has to be accepted that projects sometimes go wrong. The technical problems prove to be greater than antici-pated, and the cost of solving them begins to increase drastically. Marketplace needs may change, alternative technology may be developed and competitors have a habit of doing the unpredictable.

In these circumstances, the best course of action may be to terminate a project. Herein lies the trap. We hear statements such as 'We have already spent £5 million on this project – we can't stop now' and 'The solution is just round the corner'. It always is, but the corner seems to move further away! The most insidious reason for not wanting to terminate a project is 'We have told corporate headquarters about it, the President/Managing Director has taken a personal interest in it – there will be too much "egg on our face" if we stop now'. Managers carry on in the vain hope that the problem will go away, only to end up terminating the project at a later date when even more money has been wasted.

Money already invested in a project is a 'sunk cost' – nothing can bring it back. Unless there are sound reasons for believing that the prospects of success have significantly increased, the wise course of action may be to accept the loss and move on to something else which can be more profitable (Exercise 11.5).

EXERCISES

11.1 Analyse the business development activities of your company over, say, the last three years, in terms of the four categories in Section 11.1. How much future income (or, preferably, profit) has each one generated? What unexploited opportunities does this analysis suggest?

11.2 Estimate the marketing cost of innovation for the last financial year, and compare it with the technical cost. Are these two figures in a sensible relationship? (See also Follow-up 11.1.)

11.3 Put your current new product development projects through the five filters listed in Section 11.3. Does this lead you to change your invest-ment decisions in any way? (See also Follow-up 11.2.)

11.4 Carry out the process of 'identifying winners' described in Section 11.4. (This will involve assembling an appropriate team, and ensuring that they can devote the necessary time to the exercise; the project should first be 'sold' to senior management, so that it receives their backing.)

11.5 Ask yourself (honestly!) whether you are falling into any of the traps on impending failure described in Section 11.6?

FOLLOW-UP

11.1 For your business plan period (three to five years?), set out the estimated marketing cost of innovation beside the technical cost. Consider ways in which your innovation could become more market led, and estimate the cost of making it so. What would be the effect on long-term profitability of this changed emphasis?

11.2 Incorporate the five filters in Section 11.3 into your new product evaluation procedures.

11.3 Incorporate the 'risk and sensitivity' analysis described in Section 11.5 into your investment appraisal procedures, using discounted cash flow (see Appendix) unless you already have an equivalent procedure.

REFERENCE

Ansoff, H. I. (1988), *The New Corporate Strategy*, New York: John Wiley.

APPENDIX: DISCOUNTED CASH FLOW

Although the DCF technique was introduced many decades ago, it still represents one of the simplest aids to investment appraisal. Such investments might include building a new factory, installing a new machine or, in the context of this chapter, developing a new service or entering a new market segment.

Although it is possible to use a computer model, a spreadsheet or DCF tables, the calculation can be done with a calculator (or even with pencil and paper).

The basis of the technique can be understood by asking 'would we be willing to lend someone £100 if they promised to repay us £100 in a year's time, with a guarantee of no risk?' No, because we could put the money into a bank and earn some interest. Suppose they offered us to repay us £105, £110, £115, £120...? The point would come where we would agree to the deal because it was better than other ways of investing the money. In other words, we have a certain view of what we expect our money to earn for us. If the figure were 10 per cent, we would say that £110 in a year's time was equivalent to £100 today, or that £110 in a year's time had a *net present value* of £100.

It is important to realize that this is nothing to do with inflation, but relates to the time-value of money. Inflation is obviously a complicating factor, but it may be possible to eliminate its effect by assuming that we can raise prices in line with inflation, i.e. we do the calculation at present-day prices and assume that inflation will cancel out.

Let us now apply this thinking to an investment. We have a number of sources of cash outflow and cash inflow. If we simply added up the outflows and subtracted them from the inflows to see whether we had made a gain or a loss, we would be ignoring the time effect. The problem is that, in most cases, we are investing the valuable pounds at the start and in the early years of the project, and receiving back the less valuable pounds in the later years of the project. This gives us an artificially favourable view of the value of the investment, particularly when most of the sales come late in the time period.

Let us apply the DCF technique to the investment in Section 11.5. See Table 11.A.

Table 11.A					
	Year 1	Year 2	Year 3	Year 4	Year 5
Cash inflow (£'000)*					
Sales revenue	0	500	750	1250	1500
Total cash inflow	0	500	750	1250	1500
Cash outflow (£'000)**					
Cost of sales	0	300	450	750	900
Expenses	100	200	225	250	300
Total cash outflow	100	500	675	1000	1200
Net cash inflow	(100)	0	75	250	300
Total undiscounted cash inflow: £525 000					

Notes: * Other cash inflow might arise from increase in creditors/payables, buildings and equipment sold, royalties.
** Other cash outflow might arise from increase in debtors/receivables, buildings and equipment purchased, increase in stock (if applicable), tax.

It could be argued that the above figures do not strictly represent the entire cash flow, because of the time phasing within the year and because they exclude some of the cash items, but they will suffice to illustrate the technique. Refinement of the figures may not lead to significantly different conclusions; if other items are significant, they should of course be included.

If we simply added the figures, the cumulative cash inflow would be £525 000 and we would feel favourably towards the project.

However, if we discount the cash flows at 10 per cent, the picture is very different. To do this, we need to divide the cash flow:

in year 2 by 110 ÷ 100 = 1.1
in year 3 by 1.1 × 1.1 = 1.21
in year 4 by 1.1 × 1.1 × 1.1 = 1.33
in year 5 by 1.1 × 1.1 × 1.1 × 1.1 = 1.46

These figures can be obtained from DCF tables, but they can easily be calculated by putting 1.1 into the memory of a calculator and dividing the year's net cash inflow the appropriate number of times.

This changes the net cash inflows from:

Table 11.B

	Year 1	Year 2	Year 3	Year 4	Year 5
Undiscounted	(100)	0	75	250	300

to:

Table 11.C

	Year 1	Year 2	Year 3	Year 4	Year 5
Discounted at 10%	(100)	0	62	188	205

We can now add these *discounted* figures to give the net present value:

NPV at 10% = £355 000

This is significantly different from the original sum of £525 000 without discounting, and may cause us to revise our degree of optimism in the project.

If we were to use a discount rate of 20 per cent, the figures would be:

Table 11.D

	Year 1	Year 2	Year 3	Year 4	Year 5
Discounted at 20%	(100)	0	52	174	174

The NPV at 20 per cent is £300 000, which is little more than half of the undiscounted total.

Some might say that a discount rate of 20 per cent is much too 'greedy'; by looking for this sort of return on investment, worthwhile projects might be rejected. Against this argument, we must remember that some projects fail to reach their expected potential or even fail altogether. The 'successful' projects have to pay not only for themselves but for the unsuccessful projects as well. We may therefore have to use a higher discount rate, depending upon the risks in our particular type of business.

An alternative to using the net present value based on a predetermined annual interest rate is to calculate the interest rate which would give an NPV of zero. This rate is known as the internal rate of return (IRR). This can be an aid to ranking a number of projects which are competing for limited resources, but we must not forget the magnitude of the cash flows; a very high percentage IRR is not much good if the amount of money involved is small.

As we said at the start, much more sophisticated evaluation techniques can be used, but the DCF method can be used by anyone. Certainly no marketing managers reporting to me would ever have submitted an investment proposal without a risk and sensitivity assessment (see Section 11.5) and a DCF calculation – they knew that it would be referred back to them if they did (Follow-up 11.3).

12

HOW DO WE MANAGE THE FUTURE?
MARKET-BASED BUSINESS PLANNING

Key business issues   *Section*

O   Business planning must be based on marketplace
    realities, not on internal aspirations   12.1

O   Business groups have the advantage of defining clear
    profit responsibility for achieving results in the
    marketplace   12.2

O   There may be a role for a full time business
    development function, but qualified fee-earners are
    often the best people to market their own highly
    sophisticated services

O   Plans should be regarded as essential working
    documents, not theoretical exercises   12.3

O   The business plan sets out a viable future for the
    business and ensures that everyone is working to the
    same objective   12.4

○ It is strongly recommended that a business planning workshop should be set up

○ The marketing plan shows that marketing strategies and tactics have been clearly thought out, and ensures that everyone understands the commitment to marketing which the plan requires   12.5

○ The sales plan forces us to say who is going to do what by when   12.6

○ Internal business proposals are marketing documents. Some key factors which would lead to early approval are often ignored   12.7

○ Models and techniques can assist us in understanding our business situation   12.8

○ They are only worth using if we treat them realistically and act upon the findings

12.1 THE BASIS OF BUSINESS PLANNING

Business planning must be based on marketplace realities rather than on internal aspirations. This is yet another example of 'out-to-in' thinking which has frequently been commended in this book.

The mistake that many people make is to start with the amount of income they need to achieve – to cover the growth in their overheads, to show corporate senior management that they are a successful operation or simply to earn more money for themselves or their organization. Unless this figure is based on a realistic estimate of the amount of business which customers will give us and competitors will allow us to have, it is a dangerous delusion.

There is a very simple but fundamental formula:

Turnover = market size × market share

Market size is the total amount of money we estimate that customers will spend on the type of services we offer in a given period, normally a year. This amount will obviously depend on how we set the boundaries of our market. If we define our customer base very broadly, e.g. those wanting computer services, we will have a small share of a very large market. If we define our market very specifically, e.g. those requiring computer integrated manufacture in the motor industry, we can expect a higher share of a smaller market. It is argued in Section 4.7 that the preferred position is to dominate or at least to be a main supplier in one or more closely defined niche markets than one of many suppliers dabbling in a loosely defined general market.

Market share is the percentage of this market size which we expect to achieve. If we knew the market share of all our competitors, the total including our own share would be 100 per cent.

The importance of this method of estimating turnover can be illustrated in a number of ways.

First, it deters us from being over-optimistic. 'We only want 10 per cent of the market' – but so do the other 20 competitors! Without an unreasonably costly amount of market research, we cannot know the competitors' shares accurately, and even then the figures would be based on past performance not future; nevertheless, it is important to estimate the market shares of at least the main competitors with a reasonable degree of certainty, so that our own forecasts have some foundation in reality.

I remember when a new food product was being introduced in plastic tubs. At the time I was working in a packaging company which was being asked for quotations for plastic tubs for all the main would-be manufacturers. By adding all the estimates together, they were in the unusual position of being able to see the sales forecast for a whole industry. The total was equivalent to four tubs for breakfast, lunch and dinner, every day of the week, for every man, woman and child in the country! Yet these figures had been provided by managers who were making decisions about investing in a new market.

I blame laser printers for many of today's business problems! Planning documents are produced, illustrated by growth curves, multi-coloured bar charts, pie charts and so on, printed at 1200 dots per inch, bound in impressive covers and distributed to senior managers. They look so convincing that they must be right! The truth is, of course, that they have no more foundation than the sums on the back of the envelope on which they were originally written.

Second, it helps us to judge whether a certain level of planned growth is reasonable. Let us suppose that we are expecting 20 per cent growth over a period of, say, three years. Is the market expected to grow at 20 per cent? If so, should we perhaps be looking for 21 per cent or 22 per cent (as long as we have justification for this). Is the market static? If so, any increase in our turnover can only come from a competitor. Which competitors are we expecting to win this business from? Why will they allow us to do so? Unless we have convincing answers to these questions, we have no secure foundation for our growth figures.

To summarize, I would never submit or approve a business plan that did not reflect the market size × market share approach to turnover, with an accuracy which it is reasonable to expect in a particular situation.

12.2 ORGANIZATION OF THE BUSINESS DEVELOPMENT FUNCTION

In a large marketing-oriented company, in a field such as fast-moving consumer goods, there would be a complex hierarchy including a marketing director, marketing managers, product group managers, product managers, marketing service managers, sales managers, regional and area sales managers, sales representatives and so on. All these people would have marketing and/or selling as their full-time job.

For all but the largest high technology service organizations, such a large and complex structure would be entirely inappropriate. Furthermore, many in the industry would consider that the best people to market and sell their services are the fee-earners themselves ('seller-doers'). Because of their complexity, nobody but a specialist in the particular field of high technology can accurately understand the customers' needs or convincingly sell the supplier's skills.

In my experience of many high technology service organizations which are taking their marketing seriously, the optimum structure for the business development function is probably a combination of full-time and part-time staff. For the selling part of the operation, the advantages of the two approaches are discussed in Section 8.6. For the marketing part, while fee-earners certainly have a necessary contribution to make, the demands on their time may not allow them the uninterrupted chance to think through the complex marketing analysis and planning issues involved. One or more full-time marketing or business development staff may therefore be required, or the manager of the business unit may make it a priority in his or her own workload.

The size of a full-time business development department varies considerably in different organizations, and it is often smaller than it ought to be; the headcount for a team of several hundred fee-earners is often in single figures. Whether or not this is sufficient depends upon the extent to which the fee-earners take their marketing seriously. Also there are some activities which are best performed by a central business development function, because no individual department can adequately represent the whole range of disciplines involved. These activities include the overall company marketing strategy, handling the marketing of large multi-disciplinary projects and corporate publicity. (It is significant that corporate publicity is put last in the list, not because it is unimportant but because many people treat this as the only important role of a central marketing function. They spend large amounts of money on a new logo, a new corporate style, new notepaper etc., and ignore the basic business issues which will determine whether or not they will succeed.)

If the first question is 'how many staff should work full time in business development?', an equally important one is 'how should the fee-earners be organized?' The all too frequent answer, which is the traditional one, is 'by departments, technological speciality or skill sets'. Many companies moving towards a market-led orientation are realizing that this is not the best approach. Instead, they are reorganizing into profit centres, which may be designated business groups, business units or strategic business units (SBUs) and which reflect the way their customer base is organised ('out-to-in' yet again).

A profit centre or business group is a part of the business which, in principle, could exist as a separate company in its own right even though, in practice, it may share a number of facilities with other profit centres. Each business group has its own range of services, market segments and competitors (a unique product–market combination).

A business group is managed by a business group manager who is responsible for strategic planning and profit performance for that part of the company. In the simplest case, a business group manager is responsible for the marketing activities of the unit, normally including line responsibility for the selling function, and for line management of a number of fee-earners. Beyond this, it is quite normal for a number of business group managers to share resources such as R&D, development, personnel, buildings and so on. They often need to draw on the skills of technical specialists from other business groups, especially in multi-disciplinary projects. This obviously leads to conflicts of priority which have to be resolved between responsible managers, but the fact that something is difficult does not mean that it is wrong (any more than the fact that something is easy means that it is right).

We said initially that the business group manager would normally have direct line responsibility for the selling function. It is possible for this also to be carried out by another part of the company for which he or she is not responsible, but this can lead to enormous conflicts.

From the perspective of the customer the great merit of business group management is that they only have to deal with one person or department. Any interface problems within the service company are handled by that service company. The advantage from the perspective of the service company is that profit authority and responsibility are delegated to senior middle management who are able to concentrate exclusively on the exploitation of their particular marketplace. Business group managers are in effect 'champions', who are paid to put all their effort, enthusiasm and resources behind the achievement of a clearly defined goal for their profit centre. At a stroke, this organization overcomes many of the problems of bureaucracy, delayed decision-making and wasted expense which one often observes in more hierarchical organizations.

The two approaches to organization of the fee-earners can be represented diagrammatically, as in Figure 12.1.

The approach represented by the horizontal arrow in Figure 12.1 – the skill-set organization – tends to be 'in-to-out' and technology driven, while

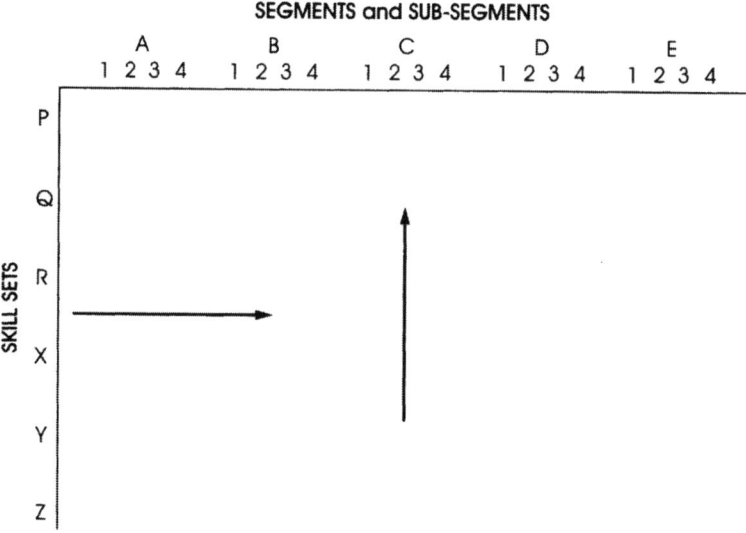

Figure 12.1   Skill-set versus market segment organization

the other – the market segment organization – is 'out-to-in' and market driven.

The skill-set approach is based on the thinking 'I have a department which needs work; I have to keep my staff gainfully employed; who can we sell to?' The market segment approach says 'there are some groups of customers out there who have business needs; how can we meet those needs and make a profit?'

There is a common situation that represents a problem for the horizontal thinker but an opportunity for the vertical. It is that a particular customer or segment does not normally require only one skill-set. For many suppliers, their greatest strength may be their ability to offer integrated solutions, where the resources come from a number of skill-sets. If the selling and marketing is left to technical department managers, each jealous for his or her own profit, the integrated solutions may never be offered.

Although it creates some problems (as does any form of matrix organiza-tion), those who have 'taken the plunge' and organized by market segments would never want to revert to a skill-based organization (Follow-up 12.1).

12.3   THE PLANNING PROCESS

Put very simply, the planning process addresses three questions:

O   Where do we want to go?
O   Where are we now?
O   How are we going to get there?

We should start by looking to the long-term future and then work back to the present. We cannot change the present situation or the process by which we have reached it, but if we are too constrained by where we are, we are in danger of simply extrapolating the past. A series of short-term tactical decisions does not constitute a strategy. We cannot assume that the organization, staff and physical resources which we happen to have at the moment are correctly balanced for the future. We may have to make some changes, thinking laterally rather than linearly.

In an organization with a mature planning process, there will be three plans, or rather one plan with three components:

1.   *A business plan*, which sets out the key strategic objectives. The time-scale for the long-range strategic part of this plan is typically three to five years, but in some industries, where the impact of investments takes much longer to be measured, a ten-year or even a 20-year time-scale may be appropriate. Within this long-range strategic plan will be a detailed plan covering the first year.

2.   *A marketing plan*, which develops those parts of the business plan that relate to the marketplace, the competition, the services being marketed and the marketing activities within the profit centre or business group. Although this marketing plan must be 'owned' by the senior marketing member of the management team of the profit centre, it has significant implications for R&D, finance, personnel and almost every other function in the organization.

3.   *A sales plan*, which details target customers, sales programmes, journey plans and so on as required to meet the marketing objectives.

Where a number of profit centres or business groups are involved, there will be a business plan for the whole organization. In principle, this should be an integration of the plans for the individual business groups. In practice, particularly if the plan is being submitted for approval to higher authority such as a corporate main board, those responsible for the combined plan may commit to a slightly less stretching target which allows for the fact that it is not realistic to assume that all of the individual targets will be achieved simultaneously (or, put less tactfully, senior managers may wish to ensure that they are more likely to beat the overall plan target than to fall short!).

We would urge readers to ensure that any plans they develop are realistic working documents. Too many companies engage in an annual ritual where the whole senior management team stops making money for several weeks and fills in bits of paper which nobody believes and no one but 'head office' reads. If the planning documents are not the almost daily guide of the line managers in the profit centre that originated them, they have not achieved a major part of their purpose.

To summarize, the aim is to produce a series of documents that are realistic and internally consistent, are approved with the minimum of disruption and are regarded as essential aids to the management process (Exercise 12.1).

12.4 THE BUSINESS PLAN

It is strongly recommended that a 'business planning workshop' is set up on the basis proposed in this section.

If the exercise is being attempted for the first time, it is reasonable that some of the detail will not be as complete as in later years, partly for lack of time (we do actually want to do some fee-earning!) but also for lack of information. We cannot stop the world while we find out everything we might want to know about the marketplace and the competition; we have to make some estimates and some judgements. However, even at a first attempt, the main points should be considered and the basic logical process followed.

In later years, particularly if the development of our business represents evolution rather than revolution, more time and information will be available to enable us to write a more detailed plan.

The process outlined in this book is based on a number of templates which are shown at the end of this chapter, on which appropriate figures and statements can be recorded for discussion, refinement and agreement.

The early sections define the key issues on which the future of our part of the operation depends. It is significant that we look first at the marketplace and only then at our resources ('out-to-in'). There is no business reason why our past mix of skills and facilities should be right for the future. We must look at the changing needs of the marketplace and then adapt our resources to match.

It is accepted that some of the numbers will be estimates or even 'guesstimates' in the early stages of the process. It is recommended that the reader puts down the best numbers and moves on to complete the process. This will reveal how significant any particular set of numbers is to the decision-making process. If a strategy depends critically on certain assumptions, more time should be spent on refining and validating the numbers on which these assumptions are based. If a closer look at the numbers would not change the basic logic, it may not matter if there is some degree of uncertainty. This somewhat pragmatic approach may seem foreign to readers trained in scientific disciplines, but we are too busy to have accuracy for its own sake.

The templates should be accompanied by appropriate narrative. (If the recommendation to hold a 'business planning workshop' is being followed, the narrative will have to be written afterwards, but main points should be recorded on the templates as they arise.) Here again we would urge pragmatism. We are not writing a book – we are aiming to present some arguments in order to persuade someone (or ourselves) to accept our plan. As long as the essential points are covered, the shorter the better.

The narrative should be definitive. Every sentence should say something – a fact that is not self-evident, a decision, an action, a problem, a solution.

Generalizations and 'motherhood' statements such as 'quality' should be avoided.

Descriptive narrative should be omitted or relegated to an appendix unless it is essential to the argument.

The whole exercise should be persuasive. We should assume (because it is normally true) that we are competing for limited funds, resources and senior management attention.

The templates are written as for one market segment. If a business unit covers more than one segment, the templates should be completed separately for each segment and the results consolidated into one overall plan for the business. This is a logical consequence of following the 'out-to-in' principle.

As we have said, a very productive exercise in the planning process is to bring senior members of the team together in a business planning workshop, off site or where they will not be interrupted. They are put into groups according to the way in which the market is segmented, and asked to go through the set of templates over a period of one or two days. The results for each group (i.e. each segment) can then be consolidated to produce an embryonic plan for the whole business. This will obviously need a great deal of refining, validating and amending, but this sort of brainstorming approach has been shown repeatedly to produce extremely valuable results, and is a way of making significant progress in an otherwise almost infinite task. In some cases, virtually the whole team has been put on the workshop; this enables everyone to contribute and feel involved, and contributes more to team-building than the sort of exercises which some companies are persuaded to adopt (Exercise 12.2).

A key to the exercise is to choose the segments with a great deal of care. The reader should refer back to Chapter 4, and consider the following points.

It is often practical to start with a twofold segmentation. Segmentation criteria might include:

O   existing and new customers
O   large and small customers
O   sophisticated and unsophisticated customers
O   customers with a continuing relationship and those placing *ad hoc* contracts
O   customers requiring a single discipline and those requiring multi-discipline services.

For businesses providing services within their own organization but also having to sell to the outside world for the first time, a segmentation between inside and outside customers can be the most useful first step.

Having chosen from the above or other segmentation criteria, it is easy to combine two of them to form four. For example, we might choose:

O   large existing customers
O   large new customers
O   small existing customers
O   small new customers.

By allocating a group of staff to each of these four segments for the purpose of the exercise, we are forced to consider the more difficult areas such as potential new customers who are unknown and might otherwise be overlooked.

The process can be further developed by a sequential segmentation. We might start with a two- (or more) fold segmentation, and then continue by sub-segmenting one (or more) of these segments. This process creates a number of segments, and we can select the most important of these for the exercise. (The less important segments should perhaps not be ignored, but they may not offer enough scope for a group to be dedicated to thinking about that segment only.)

We have said that this allocation of staff into groups is for the duration of the business planning workshop, but it has even more significance if the staff are organized in the same groups for their continuing work.

The suggested template layouts given at the end of this chapter may be photocopied for your own use. They are equally applicable if the plan is being written by individuals or in groups on a workshop as proposed above.

TEMPLATE 1A: WHAT ARE THE KEY COMMERCIAL OBJECTIVES FOR THE MARKET SEGMENT?

It is important that we should have some main commercial objectives, but we must avoid the 'self-fulfilling delusions' discussed in Section 12.1. The objectives should be specific and measurable, stretching but achievable, e.g. to achieve 20 per cent sales outside our traditional core business by the year 200–. It should be possible to monitor progress towards achieving these objectives.

It may well be that these key commercial objectives will have to be modified in the light of later templates in an iterative process, but we need to have from the start a clear idea of where we are trying to go.

TEMPLATE 1B: WHICH CUSTOMER GROUPS WILL PROVIDE THIS INCOME?

The aim is to identify those customer groups (sub-segments) within the market segment which are likely to generate the business needed to achieve our key commercial objectives. We should give specific examples of customers (but not figures at this stage). This process will begin to force us to face the realities of winning new business, and will prevent us from relying on generalizations such as 'there must be some business out there somewhere'.

TEMPLATE 2A: WHICH CURRENT KEY COMMERCIAL SERVICES ARE GOING TO PROVIDE THIS INCOME FOR THE MARKET SEGMENT?

What top three to five key commercial services does the business group currently undertake which relate to this market segment? (Possible new services will be considered in the next template.) These services should be formulated as precisely as possible without generalizations, e.g. not 'laboratory testing' but specific test programmes. A final category can summarize the 'rest'.

We should recognize that a commercial service is a business activity, not just an activity which we spend time performing. It is a service that a customer receives and pays for, not a task we do. The services listed should be made from the customers' perspective, not ours.

TEMPLATE 2B: NEW STRATEGIC INITIATIVES

In the light of the key commercial objectives (Template 1a), what new strategic initiatives are required, and what new services will result from or be required in order to take these strategic initiatives?

TEMPLATE 2C: STRATEGIES REJECTED

What main strategic business options have been considered and rejected, and what are the main reasons for their rejection? 'Do nothing' may be one of the options in some areas of the business, but this should be a considered conclusion and not one reached by default.

The reason for considering strategies rejected is that it helps to clear our minds and is much more persuasive for those who have to approve the plan. If we say 'we have considered three strategic options, rejected the first because the competition is already too entrenched, rejected the second because it is a commodity market and nobody is making any money, but chosen the third because it is a niche which we plan to dominate within three years on the basis of our unique skills', the reader will conclude that we know what we are doing. If we simply say 'we think we ought to do so and so', the reader might think otherwise!

TEMPLATE 3: WHY SHOULD THEY BUY FROM US?

What is unique, special or differentiated about the benefits we have to offer? What benefits can we offer which some competitors can also offer? What benefits are offered by competitors and not by us?

This template is an objective appraisal of where we stand in relation to the competition. This important issue can be explored in more depth in the SWOT analysis (Section 12.8).

TEMPLATE 4: WHO ARE WE UP AGAINST?

List the main sources of competition for your services in the market segment, using some or all of the five forms of competition described in Section 5.1:

1.   Other providers of the same service.
2.   Other ways of doing the same thing.
3.   Providers of an alternative service (it may not be possible to distin-
        guish this from 2 above).
4.   Other competitors for the customer's budget.
5.   The customer as competitor.

In each category, summarize the arguments you would use to overcome this particular form of competition.

TEMPLATE 5A: HOW DO WE SET MARKET-BASED PRICES?

Our aim here is to set prices that are based on value to the customer rather than cost to us. The exercise should be worked for each of the main commercial services defined in Template 2a, with a 'catch-all' for the rest.

In some cases it is possible to quantify the value fairly precisely, in terms of increased output, better yields, less waste, lower operating costs etc. In other cases this is not possible, and yet we are convinced that we are offering value to the customer, otherwise why would they use our services? In this latter case we should still refer to the factors which are linked to value, and even put an order of magnitude on this value. I argue that if a delegate attending my seminar is able to win one contract they otherwise would not have won, or to charge a higher price than they otherwise would have done, they have recovered my fee many times over. If an in-house session does not lead to benefits valued at least at tens of thousands of pounds, the event will not have succeeded.

The 'reasonable price' in column 4 should be linked to the financial value in column 3. We may decide that we could claim a certain percentage of the savings or profit improvement.

We have referred to pricing as the under-exploited management opportunity (Chapter 6). This template is a means of exploiting this opportunity. Remember that the effect of price on bottom-line profit is enormous. A difference of a few percentage points may make all the difference between profit and loss, expanding or contracting the department and, for some people, keeping or losing their job.

TEMPLATE 5B: COMPETITORS' PRICES

Here we should list the competitive prices for the services most similar to our own, and note any action needed to find out more about these competitive prices which can be a significant factor in determining what the customer feels is the 'right' price.

This action should be specific – who is going to do what by when? A generalized action 'find out about competitors' prices' is not likely to achieve much.

TEMPLATE 5C: FUTURE PRICE TRENDS

Many companies experience a dilemma at this point. They realize, perhaps, that they have been significantly under-pricing, and have a reasonable idea of what their prices should be. The difference might be substantial, perhaps 50 per cent or even more. The question is 'can we introduce such a large increase in one step, or should we phase it in over a number of years?' Any increase is resisted by customers, and repeated double figure percentage increases cause maximum adverse reaction.

Sometimes there is a 'once-for-all' change in the status of an organization, such as when an internal reactive cost-based service provider becomes a normal commercial operation. They are now required to charge commercial rates even to their own 'internal customers'. The recommended strategy is to use this once-only opportunity to 'sell' the fact that the new situation bears no relation to the old. Customers who were previously paying to support the internal organization via overheads now only pay when they are using it (and they might well pay less than before). The same would apply to any previously non-commercial operation, such as a government laboratory or a university moving into a commercial environment; there is no reason for them to start charging anything lower than the full market price. The template should show any phasing in of price increases if this route is chosen.

TEMPLATES 6A, 6B AND 6C: THE MARKET SEGMENT

The aim here is to establish the income or turnover figure that the organization will achieve in the chosen market segment. A turnover figure which is not related to market size and market share has no credibility, and is often a 'wish' rather than a planned invasion of a certain part of the market. The exercise should be undertaken for each of the key commercial services defined in Template 2a, with a 'catch-all' for the rest, and the figures added to give a total.

We use the logic (see Section 12.1):

$$\text{Turnover} = \text{market size} \times \text{market share}$$

Template 6a: Market size

Market size should, if possible, be related to the number of potential customers in the market segment who are likely to place contracts for our particular type of service. For example, we might judge that, of the 500 installations world-wide, 10 per cent of them are likely to award contracts to an average value of £200 000 in the first year of the plan, so the market size is £10 million.

It is important to realize that this is a *business plan* not a *work plan*. We must totally disregard the size of our resources when estimating the size of a market. Plans that start with the required turnover and work backwards have no foundation in marketplace reality.

The template should show the expected market size over, say, three years (or show Year 1 and an estimated annual percentage increase).

Template 6b: Market share

The pattern exactly parallels that of Template 6a, with the estimated market share being put against each market size. Where an increase in share is being predicted, we should list the competitors from whom we believe we shall win this share, and show a good reason why we think we will succeed.

Template 6c: Turnover

Again, this follows the pattern of Templates 6a and 6b, and is the product of the two for each key commercial service. This is the total income figure that will be used in the profit and loss account. The figures for the different market segments can be combined to give a total for the entire market.

TEMPLATE 7: DECISIONS AND ACTIONS

In Chapter 4, two main purposes of market segmentation were emphasized – decisions and actions. In this template we should list the main decisions which have been taken or which need to be taken for the plan to be implemented and the plan targets achieved. The main actions often follow logically from these decisions.

This exercise is a very good means of clarifying in our own minds what needs to be decided and actioned. It is also a method of involving senior management in the decision-making process. If the plan is approved, it is logical to assume that the decisions and actions have been approved. If these are controversial, the discussions during the planning approval process will automatically be focused on the main strategic issues on which achievement of the plan depends.

TEMPLATE 8: HUMAN RESOURCES ASPECTS OF THE PLAN

Do we have the right people with the right competencies and skills? Are any changes required in the balance of these competencies and skills? If increases are required, where will we source them (outside contractors, elsewhere in the organization, recruitment)?

Do we have a surplus in any area, compared with the market demand for these resources? If so, how will we reduce the surplus? We need to remember that any expense involved in changing resources will probably have to be funded from our own income.

Does the exercise identify any training needs? If so, how will they be met?

TEMPLATES 9A, 9B, 9C AND 9D: EXPENDITURE

We are now beginning to develop the profit and loss account where, of course,

$$\text{Income} - \text{expenditure} = \text{profit}$$

Different accounting conventions are used, so it would be worth ensuring that the structure of the cost centres and the allocation of different items of expenditure are consistent with the normal reporting methodology for the organization, unless they are not appropriate to the new market-based situation and need to be altered.

All items of expenditure ultimately have to be included, but we concentrate initially on the expenses involved in marketing and selling because these are likely to require the most thought.

People frequently ask me for guidelines on the percentage of turnover that should be spent on marketing and selling. My answer is 'probably between 5 per cent and 50 per cent, but it might not be'! The reason for this is that the amount of marketing needed will depend crucially on the nature of the task to be performed. If most of our income comes from existing customers in a mature market, a few per cent may be adequate. If we are entering a completely new marketplace, the marketing expenditure before the first income is generated is an infinite percentage of turnover; it will gradually fall to an acceptable level, but it could easily be in double figures per cent for the first few years.

The percentage also depends on the typical size of a contract. It normally does not take anything like ten times as much marketing to negotiate a contract for £500 000 compared with one for £50 000. It also depends on the degree of competition. If competition is intense, more marketing will be needed, and there will be a lower success rate; it is only the successful contracts that bring any income, and these have to pay the marketing costs of the failures.

If we are becoming much more 'commercial' than in the past, are venturing into new markets or are offering new services for which we are not known, my advice would be 'don't underestimate the amount of marketing which needs to be done'. Beware the fallacy that it is technical people who create wealth and marketing people who dissipate it. We started this book by saying that marketing was there to *create* profit.

Template 9a: Staff related marketing expenditure

We need to decide who is going to do the marketing and selling, and then estimate the time that this will require in terms of full-time equivalent staff. This time should be built up by estimating the amount of each type of activity, including, for example:

○   researching the market and competition
○   developing marketing strategy
○   developing and implementing marketing programmes – advertising, exhibitions etc. (but excluding the non-staff related costs which are assessed separately below)
○   selling to existing and new customers
○   preparing and submitting proposals
○   continuing customer support
○   etc.

We then use an appropriate figure for the full cost (including overheads) of a full-time equivalent person, to arrive at the total cost of staff-related marketing expenditure.

This exercise should be undertaken separately for each market segment, as the marketing requirements may differ widely between segments (e.g. mature as opposed to embryonic).

Template 9b: Non-staff related marketing expenditure

This is the cost of all marketing and selling activities not covered by salaries and overheads. It includes, for example:

○   promotional literature
○   advertising
○   travel, accommodation and subsistence
○   exhibitions
○   etc.

Template 9c: Non-marketing expenditure

For an organization that has hitherto undertaken little marketing or selling, this would be the annual cost of all the previous activities (technical, financial, administration, management etc.), both staff and non-staff related. The fallacy is, of course, that this would mean that the marketing and selling expenses are wholly incremental and would have to come out of profit. It may well be that, in a new commercial mode, some of the non-marketing activities that were previously carried out will have to be reduced or sacrificed altogether. The profit motive does concentrate the mind!

Template 9d: Total expenditure

This is the summary of the expenditure in Templates 9a, 9b and 9c.

TEMPLATE 10: FINANCIAL SUMMARY

This template brings together the income and expenditure, and gives the first indication of profit or loss.

Much of the input data may need to be refined, and it may become apparent that the business as set out is not viable. We may have to revise some of our strategies. However, we must not change the figures to give the result we want: 'I'm sure we could sell 10 per cent more than that!' As with all aspects of business planning, the figures must be based on an objective understanding of external marketplace realities. If there is bad news, we need to bring it into the open.

Once the plan is formalized, a company financial expert will need to look at the cash flow implications to ensure that they are compatible with the organization's actual or potential cash resources. Where products are being manufactured, a high growth rate can place heavy demands on cash in the form of stock and debtors (inventory and receivables), but this is less of a problem in a service organization where no stock is being built.

TEMPLATE 11: CAPITAL EXPENDITURE

This should be handled in accordance with the organization's normal accounting conventions, including funding, depreciation etc. If some of the new strategic initiatives require capital expenditure, this should be highlighted to assist in the understanding of the plan and the decision-making process.

TEMPLATE 12: CONTROLS AND MONITORING

However much effort we put into writing the business plan, it is at best a judgement at a point in time. When we look back on the period every figure is likely to have been proved wrong. If the variances largely cancel out (e.g. we sell less but are able to achieve a higher average price) most people would be happy, unless there were some serious strategic implications such as a declining market. However, in my experience, far more of the variances tend to be adverse, with the result that the planned profit is not achieved.

We therefore need a system for monitoring and control during the plan period. This must not be too cumbersome, but it must enable us to see early on in the financial year whether we are on target. Corrective actions may take months for their effects to be felt.

A sophisticated plan should include a risk and sensitivity analysis, as described in Section 11.5.

Once these templates have been completed, perhaps in a workshop as suggested, the results can be consolidated and areas of duplication or omission resolved. The same structure should be used for writing the narrative part of the business plan, and for presentation for approval.

12.5 THE MARKETING PLAN

The marketing plan defines in much more detail those aspects of the business plan that relate to the marketing function. As with the business plan, the logic is:

O   Where do we want to go? (Marketing objectives)
O   Where are we now? (Analysis of the present situation)
O   How are we going to get there? (Strategies and actions)

As with other parts of the planning process, the marketing plan is a means of ensuring that marketing strategies and tactics have been clearly thought out. Without this attention to detail, the plan simply cannot be written. The plan also ensures that everyone understands the commitment to marketing effort and expenditure which the programme requires.

A typical structure is shown below.

1. Executive summary: key issues, strategic initiatives and their implications. *Non-marketing readers can be strongly influenced by the summary.*
2. Key marketing objectives (*'where we want to go'*).
3. Key assumptions (*on which achievement of the plan depends*).
4. The present situation (*'where we are now'*) .
    a) The economic environment (*availability of money*)
    b) The marketplace (*size, shares and trends*)
    c) Company SWOT analysis (*see Section 12.8*)
    d) Competition (size, market share, strengths/weaknesses)
    e) The range of services offered
5. Market research required: this should give a brief programme for a better understanding of the marketplace and the competition, and may well involve a programme for establishing market prices.
6. Marketing strategy (*'how we are going to get there'*)
    a) Market segment strategy, including priorities and key target customers
    b) Service range strategy
    c) Pricing strategy
    d) Sales strategy
    e) Promotional strategy
    f) Literature
    g) Advertising
    h) Web sites
    i) Exhibitions
    j) Public relations
    k) Customer support strategy (e.g. setting up a help desk)
    l) Other
    These strategies will be amplified by individual action plans where appropriate.
7. Human resources implication of the plan
    a) Quantity and calibre, existing and required
    b) Recruitment
    c) Training
    d) Secondments and transfers
    e) Remuneration and incentive policy
8. Financial budgets required (*further breakdown of expenses in profit and loss account*)

A marketing plan is a very good diagnostic indicator of the degree of marketing professionalism of a company and a senior marketing team. 'If you don't know where you're going, any road will get you there!' However, all that ultimately matters is the performance rather than the plan, and the document and its accompanying action plans must define very specifically who is going to do what by when.

12.6   THE SALES PLAN

The nature of this part of the plan will depend very much on the size and diversity of the operation. If there is a large structured full-time sales force, it will describe aspects such as:

O   the structure of the department (marketing/sales director, sales manager, regional managers, area managers, sales representatives etc.)
O   allocation of responsibility for different customer groups
O   frequency of visits, journey planning
O   targets for regions, areas and individuals
O   remuneration and incentives
O   expenses (relevant parts of Templates 9a and 9b, with more specific breakdown where appropriate).

If the selling is done by the fee-earners themselves, the same issues have to be considered but probably in less detail. The important point is that each fee-earner should be clear about what is required of him or her, so that they do not leave the work to others.

12.7   SUCCESSFUL INTERNAL PROPOSALS

The principles of marketing discussed in this book can be applied to internal business proposals. One is, in effect, trying to sell an idea to senior management, usually with a view to gaining approval for spending money.

The document might be a business plan, a marketing plan, or one of the specific action plans discussed in Section 12.4. It can apply to almost any situation where the idea is not so obvious that approval is immediately and automatically granted.

Many proposals are extremely pedestrian and do not excite the reader. Attention to the following points may make all the difference between a hard-hitting proposal that is likely to be approved and a sterile process which wastes a great deal of time and still leaves senior management undecided.

EXECUTIVE SUMMARY

A good executive summary not only saves time for busy senior managers who have to approve proposals. It is also a test of clear thinking. It demonstrates that the writer has really thought through the argument. As an illustration of this, I recently heard that a major international company insists that proposals for millions of pounds of research funding must not exceed one page.

ASSUMPTIONS

If we say that a proposal is predicated on certain assumptions, the reader can challenge the assumptions as part of the decision-making process. They might relate to advances in technology, prices of subcontract services, actions of the competition, estimates of the size and rate of growth of the market and so on. Some readers of the proposal may not be able to understand the technical intricacies of the conclusions, but can legitimately challenge the assumptions on which the conclusions are based as a test of the credibility of the proposal.

ALTERNATIVE STRATEGIES REJECTED

Too many proposals fall into the category of 'take it or leave it'. A particular strategy is proposed, and the reader is invited to approve it.

However, the reader may not be certain that the suggested strategy is necessarily the best. The proposal might be referred back for reconsideration and resubmission. Other approaches might be suggested.

If the writer anticipates this by describing several alternative strategies that have been considered and rejected, the writer is, in effect, doing some of the reader's homework. This shows that the writer has made a mature business appraisal of the options and has not simply taken hold of one particular idea which happened to be attractive at the time of writing (see Section 12.4, Template 2c).

CHANGES SINCE THE LAST PLAN

Proposals, particularly annual plans, often refer to situations that have been described in earlier plans. The careless proposal writer forgets to read last year's plan, and becomes ready prey for an astute recipient who takes the trouble to do so. I recall a meeting when a senior manager with a twinkle in his eye said to one of my colleagues 'I was very interested to read your proposal, particularly since you said the opposite last year; are you sure you know what you are doing?'

There are inevitably times when circumstances change. The wise approach is to acknowledge this by saying 'Last year's proposal was based on the assumption that...; circumstances have now changed and this modifies our plans along the following lines...'. This makes the document much more credible.

A well-written internal proposal, which clearly demonstrates how new profit opportunities can be created, is a sign of professional marketing management (Follow-up 12.2).

12.8 MODELS AND TECHNIQUES

There are a number of models and techniques that can help us to understand our business situation and make better business decisions.

In my experience of using them, I have found that there is a danger that the technique becomes the objective of the process, and very little is done to make decisions as a result. Busy managers spend days 'filling in boxes' and writing down things that everyone already knows, send the results off to head office with a sigh of relief and then revert to running the business. This is a pointless exercise.

To overcome this problem, I would suggest the 'so what?' approach. Having completed the analysis, what are we going to do differently in future? What changes do we need to make in our policy, strategy or tactics? Should we change our organization or any of the managers in the organization? Should we change our competitive strategy or our service strategy?

The techniques do not, of themselves, answer any questions. What they do is help us to clarify our thinking. They make us ask questions which we might not otherwise ask, and tackle important issues which we might otherwise not consider.

The two techniques described in this book are not new, but they have stood the test of time. Sophisticated computer models are now available which may be valuable when used by specialists, but the methods described here can be applied by any experienced manager.

SWOT ANALYSIS

This exercise, as the initial letters imply, investigates strengths, weaknesses, opportunities and threats. It is normally expressed in the form of a grid (see Figure 12.2).

We start by thinking in detail about our operation, listing the strengths and weaknesses. We must be willing to record the bad news as well as the good. We should try to see ourselves as the marketplace sees us, not as we would like to see ourselves.

We then write down the opportunities open to us in the marketplace, and

Figure 12.2 SWOT analysis

the threats that might restrict those opportunities. The threats might come from economic, environmental or technical factors, or from the activities of the competition.

The strengths and weaknesses of the main competitors can be considered as part of the analysis.

It is useful to undertake this exercise with a small group of colleagues, at or around our own management level but covering a variety of management disciplines. There is a benefit in doing it off site where a more relaxed environment can be achieved and interruptions minimized. This can create a very vigorous forum where constructive argument can take place and vague generalizations can be challenged.

Having completed the analysis, argued about it and clarified the issues, we now need to ask 'so what?'. How can we use our strengths more effectively? What are we going to do about the areas of weakness that we have identified? How can we exploit the opportunities open to us? How can we overcome or minimize the effect of the threats?

If we take this practical approach to the SWOT analysis, we should be able to gain a useful input to our strategic planning without spending too much time on the exercise (Exercise 12.3).

THE GROWTH SHARE MATRIX

This technique was developed by The Boston Consulting Group, and has achieved wide acceptance. It helps us to identify the relative role and stage of development of the various services which make up our service portfolio, or of the business groups within our organization. As shown in the matrix in Figure 12.3, the grid is a plot of market growth rate against relative market share (defined as our share divided by that of the largest competitor). The position of each circle represents the position of a particular service, product range or business group, and the size of the circle indicates its monetary value.

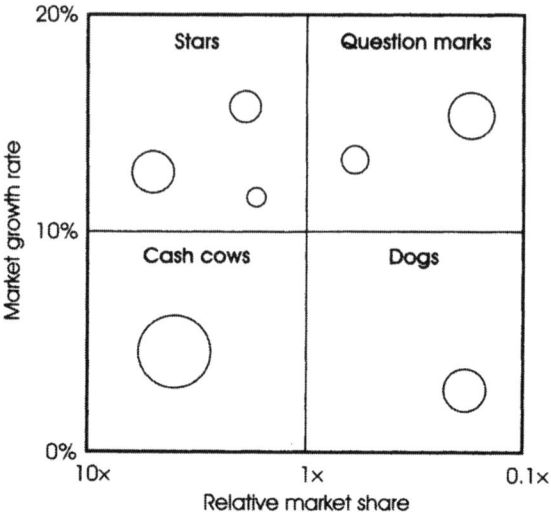

Figure 12.3 The Growth Share Matrix. Used with permission of The Boston Consulting Group. Bruce D. Henderson, 'The Product Portfolio', *Perspectives on Strategy from The Boston Consulting Group*, ed. Carl W. Stern and George Stalk, Jr. (New York: John Wiley & Sons, Inc., 1998), pp. 35–37.

As with SWOT analysis, the Growth Share Matrix is a good subject for an off-site brainstorming session as an aid to determining business strategy.

The grid in Figure 12.3 defines four situations:

1.   *Question marks.* This is a common starting point, where we have a low share in a high-growth market. A question mark has a potentially high demand on cash resources, and we may need to select the most favourable from a number of options. To summarize, invest in order to exploit the opportunity, or abandon it.

2.   *Stars.* A star has a high share in a high-growth market, and often results from a former question mark. A star may still require invest- ment of cash, but the best of the stars are the cash cows of the future.

3.   *Cash cows.* Cash cows have a high share in a low-growth or declining market. They provide cash for promoting stars, resolving question marks and divesting dogs, and are a prime source of cash for the whole company operation. As 'cows', they should be milked for as long as possible.

4.   *Dogs.* Dogs have a weak market share in a low-growth or declining market. They can cause a significant diversion of management effort and drain on profit. Some people retain dogs for sentimental reasons, and brutal steps may have to be taken.

A typical life cycle might be that a question mark becomes a star which becomes a cash cow which becomes a dog. The whole process may take years or even decades but, in many fields, life cycles are becoming progressively shorter and management actions must reflect this.

The Growth Share Matrix may throw up some possible management errors. These include:

1.   Concentrating on the short term to the detriment of the long term. Business development usually requires an investment outflow before it can generate a profit inflow, and a short-term management focus may prevent some activities from becoming stars and cash cows because they are starved of that initial investment.

2.   Expecting all business groups to have the same growth rate or return on investment. Marketplace realities simply do not allow this.

3.   Weakening cash cows by taking too much cash out of them. If we do not feed cows, they will stop producing milk!

4.   Allowing cash cows to become complacent by leaving too much cash in them. Cows should not be allowed to become fat and lazy. (Nor should business group managers!)

5.   Investing unrealistically in dogs in the vain hope of rescuing them. Our normal priority should be to reinforce success rather than to prop up failure.

6.   Maintaining too many question marks and under-funding each. A question mark is not a stable situation; we should either invest to move towards niche dominance, or else drop or divest the activity concerned.

Intelligently applied, the Growth Share Matrix can be a productive way of stepping back from our daily activities and asking ourselves where we are going (Exercise 12.4).

EXERCISES

12.1 What improvements can you make to your planning process in the light of Sections 12.3 to 12.6?

12.2 Strongly consider setting up a business planning workshop as suggested in Section 12.4.

12.3 Carry out a SWOT analysis on your company and on your main competitors, as suggested in Section 12.8. What are you going to do with your findings? Incorporate the technique into your annual planning procedure.

12.4 Complete the Growth Share Matrix as described in Section 12.8. Consider the six 'management errors' listed. What new inputs to your strategic planning does this exercise suggest?

FOLLOW-UP

12.1 If you do not already have them, consider the implications of setting up market segment based Business Groups in your organization.

12.2 Start using the guidelines in Section 12.7 to prepare internal proposals.

Template 1a

KEY COMMERCIAL OBJECTIVES
1.
2.
3.

Template 1b

CUSTOMER GROUPS (sub-segments)	EXAMPLES OF CUSTOMERS
1.	
2.	
3.	

Reproduced from *Marketing High Technology Services*, Colin Sowter, Gower, Aldershot

Template 2a

CURRENT KEY COMMERCIAL SERVICES FOR THE MARKET SEGMENT
1.
2.
3.
Rest

Template 2b

STRATEGIC INITIATIVES	NEW COMMERCIAL SERVICES
1.	
2.	
3.	

Reproduced from *Marketing High Technology Services,* Colin Sowter, Gower, Aldershot

Template 2c

STRATEGIES REJECTED	REASONS FOR REJECTION
1.	
2.	
3.	

Template 3

UNIQUE / SPECIAL / DIFFERENTIATED BENEFITS

BENEFITS OFFERED BY US AND SOME COMPETITORS

BENEFITS OFFERED BY COMPETITORS, NOT BY US

Reproduced from *Marketing High Technology Services*, Colin Sowter, Gower, Aldershot

Template 4

OTHER PROVIDERS OF THE SAME SERVICE	
SOURCE OF COMPETITION	ARGUMENT(S)

OTHER WAYS OF DOING THE SAME THING	
SOURCE OF COMPETITION	ARGUMENT(S)

PROVIDERS OF AN ALTERNATIVE SERVICE	
SOURCE OF COMPETITION	ARGUMENT(S)

OTHER COMPETITORS FOR THE CUSTOMER'S BUDGET
ARGUMENT(S)

THE CUSTOMER AS COMPETITOR
ARGUMENT(S)

Template 5a

	SERVICE	BENEFIT OFFERED	VALUE TO CUSTOMER	REASONABLE PRICE
1.				
2.				
3.				
Rest				

Template 5b

COMPETITOR	PRICE	ACTION NEEDED (by whom by when?)

Reproduced from *Marketing High Technology Services*, Colin Sowter, Gower, Aldershot

Template 5c

SERVICE	PRICE			
	CURRENT	Year 1	Year 2	Year 3 .....
1.				
2.				
3.				

Template 6a

SIZE OF MARKET SEGMENT		
SERVICE	YEAR 1 (£)	YEAR 2 (£) YEAR 3 (£) or FUTURE TREND % P.A.
1.		
2.		
3.		
Rest		

TOTAL SIZE OF MARKET SEGMENT (YEAR 1) = £.........................

Reproduced from *Marketing High Technology Services*, Colin Sowter, Gower, Aldershot

Template 6b

YOUR SHARE OF THE MARKET SEGMENT %		
SERVICE	YEAR 1 (%)	YEAR 2 (%) YEAR 3 (%) or FUTURE TREND % P.A.
1.		
2.		
3.		
Rest		

FROM WHOM WILL YOU WIN THIS MARKET SHARE, AND WHY?

Template 6c

TURNOVER IN MARKET SEGMENT		
SERVICE	YEAR 1 TURNOVER (£)	YEAR 2 (£) YEAR 3 (£) or FUTURE TREND % P.A.
1.		
2.		
3.		
Rest		

TOTAL SEGMENT TURNOVER (YEAR 1) = £ ....

Reproduced from *Marketing High Technology Services*, Colin Sowter, Gower, Aldershot

Template 7

KEY DECISIONS		
DECISION	DATE	BY WHOM

KEY ACTIONS		
ACTION	DATE	BY WHOM

Template 8

ADDITIONAL RESOURCES REQUIRED	
COMPETENCIES AND SKILLS	SOURCE

RESOURCES NO LONGER REQUIRED

TRAINING NEEDS

Reproduced from *Marketing High Technology Services*, Colin Sowter, Gower, Aldershot

Template 9a

STAFF RELATED MARKETING EXPENDITURE
Selling / marketing done by .... This requires .... full time equivalent people Cost @ £................. per full time equivalent person per annum = £.......................

Template 9b

NON-STAFF RELATED MARKETING EXPENDITURE	
Literature / advertising	£................
Travel / accommodation / subsistence	£................
Exhibitions	£................
Other	£................
TOTAL NON-STAFF RELATED MARKETING EXPENDITURE	£....................

Template 9c

NON-MARKETING EXPENDITURE	
Staff related expenditure	£....................
Non-staff related expenditure	£....................
TOTAL NON-MARKETING EXPENDITURE	£....................

Reproduced from *Marketing High Technology Services*, Colin Sowter, Gower, Aldershot

Template 9d

TOTAL EXPENDITURE	
Staff related marketing expenditure (from 9a)	£ ....
Non-staff related marketing expenditure (from 9b)	£ ....
Non-marketing expenditure (from 9c)	£ ....
TOTAL EXPENDITURE	£ ....

Template 10

FINANCIAL SUMMARY FOR SEGMENT FOR YEAR 1	
Total income (i.e. turnover) (from 6c)	£ ....
Total expenditure (from 9d)	£ ....
PROFIT or SURPLUS (income – expenditure)	£ ....

Template 11

NEW CAPITAL EXPENDITURE
(set out in accordance with the organization's normal accounting conventions)

Template 12

KEY INDICATORS OF PERFORMANCE

OTHER METHODS OF MONITORING PROGRESS

MAIN RISKS

CONTINGENCY PLANS

Reproduced from *Marketing High Technology Services*, Colin Sowter, Gower, Aldershot

INDEX

Advertising 9, 19, 49, 72, 116, 122, 125,
    164–6, 169, 184, 220, 221, 223
  agency 125, 165
Agent 162, 185
Authority level 104, 133

Benefits 8, 32–7, 39, 72, 86, 154, 233, 235
  dis-benefits 35
Brand, branding 39, 44
  image 40, 108
  name 40
  personality 39
Brochures *see* Promotional literature
Budget 85, 104, 125, 165, 180, 223
Buying decision 28, 45, 68
Business
  decisions 174
  definition 64, 81
  development 66, 190–92, 208–10
  group 209, 215, 230
  management 12
  models 226-9
  plan 197, 207, 211, 212–22, 224, 230
  planning 79, 80, 205–41

Capital expenditure 222
Closing the sale 134-6, 141
Commodity 101–102, 111
Communications 72, 115–70
  impersonal 121–6, 143–70

personal 121–6, 127–41
  process 116–17, 121, 165
  strategy 123
Company name 79
Competition 31, 75, 83–91, 161, 162, 216,
    217, 220, 223, 227, 233, 234
  analysis of 85–7
Competitive
  information 87–8
  literature 146
  prices 103–104
  services 88
  situation 84–5, 149, 150,
  tactics 85, 88-91
Concept testing 196
Core business 76
Corporate identity 39, 168
Cost 16, 110, 198
  centre 6
  -plus pricing 94–5, 97–9
  sunk 200
  'true cost' 100–101
Culture 7, 51–6
Customer 68, 87, 124, 183, 231
  as competitor 85, 234
  existing 138–9, 141
  loyalty 134
  needs 6, 8–10, 19, 117, 120, 130, 131,
    137

orientation 12, 62, 146, 150
service 19
CVs 151

Decision-making group (DMG) 36–7, 44, 146, 150, 160
Demand 7, 10, 42, 121
Differentiation 28, 38, 89, 164, 192, 233
Discounted cash flow (DCF) 201–204
Diversification 190, 191–2
matrix 190

E-mail 9, 153–5
Exhibitions 88, 158–64, 170, 185, 220, 221, 223
advantages 159–60
disadvantages 160–61
follow-up 163–4
invitations 163
objectives 161-2
opening questions 163
training for 159

Features 39, 86
and benefits 32–5, 44
Fee-earning staff 18, 87, 136–8, 209
Finance,
financial 12, 86, 151, 221, 223, 240
-led 50
Free market 89, 93, 95, 99

Growth Share Matrix 227–30

Image 40, 133, 147, 151, 162, 177, 182
Industry leader 89–91
Innovation 189–204
role of marketing in 192, 200
Inserts 121, 166
Intangibility 30, 40–41, 44, 133, 137
International marketing 12, 139–41
Investment 50, 54, 71, 100, 200, 202, 228

Leads 119, 159, 161
Letters 36, 72, 116, 153–7, 169
reply form 155–7, 170
Lifetime cost of ownership 28
Literature see Promotional literature, Brochures

Mail shots 153–7
Management 194
style 78, 174-5

Margin 15–16, 100, 110
Market
definition 64
development 191
key 140
-led 19, 21, 47–62, 201
saturated 107
segmentation see Segmentation
share 14–15, 86, 206–207, 218, 237
size 15, 52, 206–207
Market research 6, 88, 162, 173–87, 220, 223
primary and secondary data 175
qualitative 176–9
quantitative 179–80
sampling 176
see also Marketing research
Marketing 4, 16, 18–19, 79, 194
budget 5, 79
communications see Communications
department 6, 10–12, 20, 55, 57, 79, 193, 208–209
expenditure 219–21, 239
function 56-8
management 17
mix 25–7, 43
organization 70
plan 197, 211, 222–4
professional(ism) 5, 12, 54, 55, 101, 224
services 17
staff 17, 20, 21
strategy see Strategic marketing
Marketing research 183–6
see also Market research
Media 73, 117
Message 131

New markets 190–91
Niche market 746

Objections 132–4, 141
One-stop shop 34
'Out-to-in'/'in-to-out' thinking xiii, 8, 9, 174, 206, 209, 210, 212, 213

Partner(ship) 27, 139–41, 162
Perception 26
Place (in Marketing Mix) 26
Policy 194
Presentation 26, 72, 130–32

Press releases 167–8
Price 16, 25, 28, 42–3, 132, 185–6, 235, 236
    competitive 86
    elasticity 102, 106
    guide-price 104
    the market will bear 96, 102–105
    premium 37, 55
    segmentation 71
    shortfall 198
    trends 217, 236
    -volume relationship 105–109
Pricing 50, 79, 80, 93–111
    cost-plus 79, 94–5, 97–9
    decisions 94, 110, 111
    market-based 94–7, 102–105, 216
    strategic 105, 223
    tactical 105
    under- 77, 94
    value 94–7
Proactive marketing 7–8, 53–5
Product 25
    segmentation 74–5
Profit 4, 5, 10, 20, 50, 53, 58-9, 65, 96, 99–100, 107, 197, 199, 200, 219
    centre 6, 209
Promotion 26
Promotional literature 9, 20, 36, 49, 72, 116, 122, 125, 144–8, 153, 169, 185, 221, 223
    guidelines 145–8
    market oriented 148
    multi-purpose 148
    product oriented 148
Proposals 36, 49, 149–52, 169, 220
    internal 224–6, 230
Public relations (PR) 19, 166–9, 223

Questions 104, 129–30, 163
    open-ended 129, 141

Reactive marketing 11, 51–2
Recruitment 13, 78, 88
Research and development (R&D) 56, 193
    department 191
Resources 58–62, 74, 213, 238
Risk 43, 50, 241
    and sensitivity 197–9, 201, 222
    in market research 179, 180
    -taking 78, 79

Sales
    forecast 197
    -led 49
    staff 17, 20, 87, 128, 136–8, 178
    plan 211, 224
    shortfall 198
    strategy 223
    support 73
Segment, segmentation 44, 63-182, 108, 140, 146, 160, 177, 210, 213–14, 218, 219, 236
    criteria 68–70, 213–14
    customer 74–8, 215
    industry 67–8
    product 74
    sequential 69–70, 214
Selling 12, 16–18, 45, 122, 127–41, 184–5
    the next step 117–21, 131, 159
Sharper cutting edge 38–9
Short list 7, 77, 118, 146, 149, 160
Strategic
    alliance 139–41
    decisions 70–72, 81, 238
    initiatives 215, 232
    marketing 17, 73, 208, 220, 223
    objectives 231
    options 216
    plan see Business plan
Strategic business unit (SBU) 208-210
SWOT analysis 86, 223, 226–7, 230

Targets/targeting 38–9, 59–62, 66, 73–4, 81, 117, 130, 162, 211
Technically-led 49
Telephone 116, 157–8, 170
Tender 28, 36, 49
    invitation to 120, 149–52
Test market 160, 186, 196–7
Track record 76–8, 81, 133

Uniqueness 58, 146, 151, 216, 233
    unique selling proposition (USP) 37–9, 44

Value 28, 33, 103, 110, 235
    added 58
    perceived 28
    pricing 94–7

Web sites 9, 49, 72, 152–3, 169, 223
'Win-themes' 150, 152, 170

9781138725362